Inventor's Guide to

Successful Patent Applications

Inventor's Guide to

Successful Patent Applications

Thomas E. DeForest

TAB | **TAB BOOKS**
Blue Ridge Summit, PA

FIRST EDITION
FOURTH PRINTING

© 1988 by **TAB Books**.
TAB Books is a division of McGraw-Hill, Inc.

Library of Congress Cataloging-in-Publication Data

DeForest, Thomas E.
    Inventor's guide to successful patent applications / by Thomas E.
DeForest.
        p.   cm.
    Includes index.
    ISBN 0-8306-9335-1 (pbk.)
    1. Inventions—United States.   2. Patents—United States.
I. Title.
T339.D38   1988
608.73—dc19   88-12233
                                                   CIP

TAB Books offers software for sale. For information and a catalog, please contact
TAB Software Department, Blue Ridge Summit, PA 17294-0850.

Questions regarding the content of this book should be addressed to:

Reader Inquiry Branch
TAB Books
Blue Ridge Summit, PA 17294-0850

Table of Contents

Inventor's Guide to

Successful Patent Applications

Introduction

This book teaches you how to write and file a patent application. Patent applications are filed, for the most part, by patent lawyers. It is not easy. However, with proper guidance, it is possible for a "non-lawyer" to do it, just as it is possible for a "non-carpenter" to build a garage. With a great deal of effort and some luck, you can build a passable garage—or get a decent patent—the first time. However, the best results come only after a lot of study and experience.

So, should you file for a patent without a lawyer? In many cases, you may have no choice. Patent lawyers typically charge $2000 or more to file a patent application. If you are a prolific inventor or simply do not have the money, it might be do-it-yourself or nothing. If this is the case, then go for it!

This book introduces you to the art of patent application writing. It gives you the information you need to get an application filed and through the patent office. This also serves as an introduction to the more detailed reference works listed in Appendix D; it is recommended that you follow your study of this book with reading the U.S. Government publications listed there.

CHAPTER 1

How to Protect Valuable Ideas

If an idea can be ripped off, it will be.

If you have ever been involved with a business that depended upon a new idea or product, you have faced an ugly reality: jealous competitors want to steal the idea. If the idea looks even minimally promising, competitors will materialize within weeks or even days of your entry into the market. If you have not managed to keep quiet regarding your plans, competitors could beat you into the market.

However, competition from strangers is often the least of your problems. You told potential financial backers about your business plans when requesting help with the financing. But after the financial backers have your idea, what do they need you for? Why not dump you from the project and thereby increase their profit margin?

Why indeed. Unless you make the correct legal moves to prevent your idea from being ripped off, it will be. But what legal protection is available for an idea?

In order to encourage the creation and development of useful ideas, a branch of law has developed that allows certain types of ideas to be protected as items of property: this is the law of *intellectual property*. The types of protection available include the following: (1) trade secret protection; (2) copyrights; (3) trademarks; and (4) patents. Because it is often possible to get more than one type of protection for any given idea, you should always consider how each type can apply to your situation. By taking advantage of these types of protection, you will be able to gain control over the use of your ideas even after people in whom you confide, or others, learn about them.

This chapter describes various forms of legal protection available for ideas. It also describes how several of these forms of protection can be combined for maximum defensive effect. Later chapters describe how to obtain what is often the most powerful form of protection—that provided by a patent.

TRADE SECRETS

It is simple to obtain trade secret protection: just keep the idea secret. The formula for Coca Cola is an example of a trade secret. Only a handful of people who work for the Coca Cola company know the recipe for Coca Cola syrup. The company requires that all of these people sign non-disclosure agreements in which they agree never to disclose the formula. Because the syrup is a complicated blend of ingredients that is processed in a complicated way, it is supposedly not possible for chemists to determine how syrup is made by merely analyzing it. Thus, the idea of how to make Coca Cola syrup will remain safe from competitors, unless an employee breaks his non-disclosure agreement.

But what prevents an employee from breaking his agreement? Answer: Courts protect the trade secrets of owners who have acted to protect themselves. If an owner of a trade secret protects his secrets with reasonable physical security measures and non-disclosure agreements, courts assist him by awarding money damages against those who breach the physical security or break the agreement. Courts also issue orders preventing others from using others' trade secrets.

No special forms or magic legal procedures are required to create trade secret protection. Trade secrets arise from the mere existence of secrets, plus reasonable efforts by the owner to maintain secrecy by means of physical security and non-disclosure agreements. It is best that non-disclosure agreements be in writing. Sample agreements can be found in form books available at most county law libraries. Just ask the reference librarian.

Advantages of Trade Secret Protection

A big advantage of trade secret protection as opposed to patent protection is that any idea can be protected. No government agency (such as the Patent Office) needs to decide whether your idea is sufficiently new or useful or otherwise worthy before trade secret protection can apply. For example, trade secret law protects Colonel Sander's "11 secret herbs and spices" recipe from theft by employees who make it. Patent protection probably would not be available for a recipe. The government does not consider a mere method of cooking ordinary food with ordinary spices to be worthy of patent protection. However, such a chicken recipe can be protected as a trade secret. Even non-technological information such as customer lists or sales figures can be protected.

A further advantage of trade secret protection as opposed to patent protection is that there is no limit on how long trade secret protection can last. Your trade secret will last forever, provided you maintain proper security and non-disclosure agreements. A patent lasts only 17 years.

A major disadvantage of trade secrets is this: if some competitor, acting independently, discovers the secret for himself, then he can use it as desired. Furthermore, if the product or idea is one that cannot be marketed without the secret being obvious to all who investigate the product, trade secret protection is not possible. Thus, for a product such as a transistor radio, wherein competitors can see how the product works by merely looking at the circuit board, trade secret protection will not work. For a device or product that cannot be marketed without the secret being revealed, patent protection is probably most appropriate.

COPYRIGHT PROTECTION

Work of authorship such as books or computer programs can be protected with copyrights.

How to Get a Copyright

Obtaining a copyright is simple: on the first page or screen of the work, display the © symbol, the year, and the author's name or the name of the company owning the copyright, as in:

© 1990 JOHN W. AUTHOR

If the program is burned into a semiconductor memory, write the copyright notice on a sticker and glue it onto the chip. If possible, a notice should be embedded inside the chip in ASCII code. Also, avoid writing copyright symbols with just parenthesis if possible, as in (C). A proper copyright symbol has a circle fully enclosing the letter C, as in ©.

That's all there is to it! You need not file anything with the government, but you can *register* your copyright with the Library of Congress Copyright Office. The purpose of registration is not to create a copyright, but rather to greatly increase the amount of damages that will be due from any pirates when you sue them for copyright infringement. Obtain the proper copyright form from your local reference librarian, fill it out, and send it to the Copyright Office together with the $10 registration fee. With the form, you must also send a copy of your work of authorship (usually). The details of obtaining copyright registration are beyond the scope of this book.

Limitations of Copyright Protection

The good news is that a copyright is easy to get. The bad news is that the protection of a copyright is limited. A copyright prevents others from copying the work, but it does not stop them from taking the ideas in your work and putting these ideas to use. Courts say that a copyright protects the expression of ideas rather than the ideas themselves. Thus, if you write a computer program that predicts which horse will win at the track, a copyright will prevent others from making copies of your computer code and selling these copies. However, a copyright will not prevent others from figuring out how your code works and coding new computer programs that operate according to the same principle. For this, you need a patent!

Courts have recently taken a very broad view concerning the extent to which computer programs can be protected by copyright. A copyright may cover more than just the actual code of a computer program. It can also cover the screen layouts and graphics of the program. A copyright can also cover microcode instructions inside a microprocessor chip. It might not be long before courts hold that programmable logic devices are also subject to copyright protection. Thus, it is prudent to put a copyright notice on every paper, computer program, and programmable chip that you produce. It costs little to provide yourself with copyright protection.

TRADEMARK PROTECTION

A trademark or service mark is a logo or phrase used to identify goods or services from a particular source. The phrase "Coca Cola" is perhaps the most valuable trademark of all time. A trademark does not prevent others from copying a product. It merely prevents them from calling their products by a name or labeling it with a symbol that is confusingly similar to that of the trademarked product.

How to Get Trademark Protection

You get trademark protection by taking two steps. First, select a name or symbol for the product that is "protectable." Second, attach labels bearing this mark to the product and market at least a token number of the products. It is preferable that this marketing be done across a state line so that federal law will apply. Trademark protection will arise automatically upon this use of the mark in commerce.

A *protectable mark* is a phrase or symbol that is not already in use by someone else and which is not (1) merely geographically descriptive, (2) merely descriptive of the type of goods, (3) merely a surname, or (4) likely to deceive others. In addition, a mark that looks or sounds like an existing mark is not protectable because it violates someone else's trademark rights. For example, you couldn't use a mark such as "Koka Kola" for long without becoming the defendant in an infringement suit.

Marks such as "Bronx Computer Products" or "Seattle Book Store" are merely geographically descriptive. The law does not favor geographically descriptive marks because it seems unfair for one person to be able to monopolize the name of an entire geographical area. Courts are less likely to enforce such a mark against competitors who copy it.

Likewise, names such as "The Computer Store" or "Computer Programs Inc." are not favored because they are merely descriptive of the goods or service. The law does not readily allow one person to prevent all others from using such descriptive phrases, because this would prevent others from properly describing their businesses in advertising. People often attempt to get around these rules by changing the spelling of words, as in "Kommputer Store." This probably will not work; in trademark cases, courts tend to apply this rule: if it sounds the same, it is the same.

Similarly, a mark that is primarily a surname such as "Jones Computer Store" should also be avoided. Other people named Jones would like to name businesses after themselves. The law does not easily allow one Jones to prevent the others from doing this.

Although it is not easy to get, protection *can* be obtained for surname marks, for example "Ford Motor Company" or "Hughes Aircraft." However, before the mark is protectable, the owner must prove that the average customer in the market associates the name with a particular company. You obviously can't prove this if the company is just starting in business. Thus, new businesses should avoid surname marks.

Similarly, although it is not easy, protection can be had for marks which are geographically descriptive or merely descriptive of the business if special proof is provided concerning association of the mark in the public mind with a particular company. A mark such as "Minnesota Mining and Manufacturing Company" would be an awful choice for a new business because it is geographically descriptive and, perhaps, descriptive of the company's products and services. However, because 3M Company is so well known and the average person associates the mark with the particular company, the mark is protectable. But new businesses should not be deceived by the successful use of a mark such as this one. Avoid descriptive marks!

Names that suggest something false about the product are also non-protectable, because the law prohibits deceptive marks. A mark such as "Ameri-Made" for computer devices made in Singapore is an example.

In general, you can't go wrong with a mark that is arbitrary or fanciful. Good examples of fanciful or arbitrary marks are "Apple" or "Atari" for computers or "Chicken of the Sea" for tuna.

Registration of Trademarks

Once you have decided on a protectable trademark and actually used it, at least in a token way in connection with sales of a product or service, you can have the mark registered with the U.S. Patent Office. Without an official filing, your mark is protectable only in geographical areas where it has actually been used. However, if you file with the Patent Office, you reserve the right to extend your market area nationwide without interference from competitors who started using the mark after your filing date.

Trademark protection lasts for as long as you keep your goods in the market. This can be forever. However, it only prevents others from copying the markings which you use to label your goods. It does nothing to give you a monopoly over sales of the goods themselves. For this, you need a patent!

PATENTS

If the idea is for a machine, manufacture, composition of matter, or method, or for an improvement or new use for one of the foregoing, then it is an invention that could be patentable.

A patent is a document issued by the U.S. Government to an inventor, with copies circulated to the public. It describes the invention and tells how to make and use it. It identifies the inventor and defines his exclusive rights. In suits by the owner against people who are alleged to have violated the patent, the judge looks to the exact wording of the patent to determine if the patent has indeed been infringed. What a deed is to land, a patent is to an invention.

The owner of a patent is given a legal monopoly on his invention. He can prevent others from making or using the patented invention for 17 years from the date the patent is granted, just as a landowner can prevent others from coming onto his land. With a copyright or trade secret, you run the risk that someone else will duplicate your idea independently, that is, simply without copying it or violating physical security or non-disclosure agreements. However, a patent prevents another from using your idea even if the other person invented it entirely on his/her own.

Since it is a powerful form of protection, a patent is granted only if special requirements are met. A determination regarding whether these requirements are met, and thus whether a patent should be granted, is made by the United States Patent and Trademark Office (or PTO for short). Exactly what these requirements are and how you can convince the Patent Office that they have been met is a subject for later chapters in this book.

The Federal Government grants patents to encourage and reward the creation of new technology. This is the deal: if you invent something new and agree to tell everyone how to make and use it, then you get monopoly rights for 17 years. During this time, you can sue anyone who makes or uses the invention without your permission, and (hopefully) collect money from people who do have your permission. However, after 17 years, everyone will be able to make and use the invention as they please without further interference from you.

A patent gives you the right to exclude others. It does not necessarily give you the right to make and sell the invention. For example, if you invent a hallucinogenic drug and get a patent for it, you can still be jailed for selling it at the local junior high school. You failed to get permission from the Food and Drug Administration. Furthermore, if you get a patent for an improvement on a patented invention owned by someone else, you can't use your invention without first making a deal with the owner of the underlying patent. Of course, because others can't use your patented improvement without your permission, there is a basis for a deal.

USING MORE THAN ONE FORM OF PROTECTION

A patent does not necessarily prevent you from also taking advantage of trade secrets, copyrights, or trademarks, but it can limit these other forms of protection.

Combining Patents and Trade Secrets

Because a patent is a publicly available printed document that describes how to make and use the invention, trade secret protection is not possible once a patent has issued because the secret is out. Furthermore, if you were not totally candid with the Patent Office and did not describe in your patent application the "best" way to make and use your invention, your patent is liable to be invalidated. You can't cover up what should be disclosed in the patent. You must keep your end of the bargain with the government.

Despite this duty to tell all, trade secret protection has a role to play even when you seek patent protection. Before the patent is issued from the Patent Office, your application is kept secret. Thus, trade secret protection can be maintained during this time. If the government decides not to grant a patent, your patent application remains secret forever and trade secret protection is not affected. Therefore, you can try for a patent without giving up trade secrecy as a backup form of protection.

Patents and Copyrights

Patent protection interacts with copyright protection in an interesting way. Suppose you write a computer program that does something new. You can copyright the program by merely slapping on a copyright notice. However, if the program is based upon an idea that meets the requirements for a patent, you can also get patent protection.

When applying for a patent on your program, you are required to submit written materials explaining how to make and use the invention. The invention is the idea that is behind the operation of the computer program. The best description for this purpose is a copy of the computer program itself. Thus, when your patent is issued and printed copies are made public, the computer program will also become public. (NOTE: Be sure to include a copyright notice on the program listing you file with your patent application!)

What happens to your program at the end of 17 years when the patent expires? Under the patent law, the public has a right to use everything that is in your patent. However, a copyright lasts for at least 75 years. Can you still use a copyright to prevent the public from using your computer program?

This is a difficult question. However, you should assume that a copyright on the exact computer code in your patent will not be valid after the patent expires. Of course, copyrights on other programs you have written that use the idea of the patent, but with different computer code, will still be good. Perhaps you shouldn't be worrying about what happens to 17-year-old software anyway—it will probably be obsolete.

A tragic example of software that could have been protected with a patent, but was not, is VISICALC, the spreadsheet program. The authors of VISICALC copyrighted their software but failed to seek a patent on it. This meant that they prevented competitors from copying the actual computer code and, perhaps, some aesthetic features of the program. However, the copyright did not and could not prevent competition from copying the *idea* of a spreadsheet program. Copyright protects the expression of ideas, not the ideas themselves. This hole in the defenses allowed competitors to market spreadsheet programs.

A patent could have prevented this unfortunate result for the authors of VISICALC. Because a patent can protect actual ideas, protection could have been obtained for the idea of a spreadsheet program, no matter how it was coded or expressed. Despite the fact that competing spreadsheets were written and coded by programmers working independently of VISICALC, a patent could have prevented these from going on the market. Thus, failure to get patent protection can be costly indeed.

Combining Patent and Trademark Protection

Patent protection can also interact with trademark protection. A famous example involves "shredded wheat" breakfast cereal. The inventors of this cereal filed for and were granted a patent in which they called their invention "shredded wheat." After they got their patent, they marketed their cereal in packages that were labeled with the words "shredded wheat" as a trademark.

After the patent on shredded wheat expired, a competitor began selling the same cereal in packages also labeled "shredded wheat." The inventors sued the competitor to stop this on the theory that the phrase "shredded wheat" was their trademark. The patent had expired, they admitted, but the trademark had not. Trademarks last indefinitely, so long as trademarked goods remain in commerce.

However, the court held that because the inventors had called their invention "shredded wheat" in the patent application, they could not monopolize this name after the patent had expired. The name of an invention is part of the invention, and, after 17 years, the public has a right to both the name of the invention and the invention itself. It is all part of the deal the inventor made with the government.

This is the lesson: When you file for a patent, do not give your invention a name that you would later like to use for a trademark. Do what Searle company has done with its Nutrasweet product. The name for the sweetening chemical in Nutrasweet is *aspartame*, which is the name used for the invention in the patent. However, for trademark purposes, the fanciful name Nutrasweet is used on the labels of marketed products. Using dual names allows Searte's trademark and goodwill to survive after patent expiration.

CHAPTER 2

How to Read a Patent

Before you write a patent application, you should know what a patent looks like. Refer to Fig. 2-1, which is a copy of a U.S. patent. Various parts of the application are marked with numbers in brackets, such as [22].

On page 1 of the patent, at [75], the inventors' names are printed. The inventors in this case are Klaus Affend; Axel Mohnhaupt; and Frank Jorn. These inventors are also referred to as "Affend, et. al," which is Latin for Affend and others.

At [73], the name of the assignee is given. The assignee is the company or person who has been assigned ownership of the patent rights. Thoratec Laboratories has somehow obtained the patent rights from the inventors, perhaps in accordance with an employment contract. The assignment of patent rights, as is required, was registered with the Patent Office.

The date [22] is when the patent application was filed in the Patent Office.

The patent number and the date the patent was issued are given at [11] and [45]. U.S. Patents are given consecutive numbers as they are issued. Over 4.5 million have been issued since the numbering system started in 1836. The issue date determines when the 17-year clock on patent rights begins to run. Before this clock starts, and after the 17 years have passed, no patent monopoly exists.

The numbers at [51], [52], and [58] indicate the type of technology contained in the patent. Patents are classified according to technology just as books in a library are given numbers according to subject matter. Bracket [51] gives the classification numbers according to the European Patent Office system.

[52] gives the U.S. classification numbers: 623/3 means that this patent is in class 623/subclass 3. All patents pertaining to artificial heart valves are in class 623. The numbers 91/28 shown after the semicolon are for a secondary classification that indicates the patent may reasonably be classified into some other area of technology. At the Patent Office, copies of all patents are kept in boxes called "shoes" according to class and subclass numbers. Patents and other technical literature pertaining to a particular class and subclass, such as artificial hearts with dual valves, can be found in a particular shoe in the public search room of the Patent Office, located at Crystal Plaza, Arlington, VA.

Before issuing a patent, the examiner reviews old patents in the same area of technology to verify that the application contains a new idea. The numbers by [58] tell what classes or shoes of patents were checked. It can be seen from [56] "References Cited" that the Patent Office found some old patents that it considered similar, including one from Dr. Jarvik. The Patent Office "cited" these references against the application, i.e., they wrote a letter to the patent applicant and asked for an explanation regarding how the applicant's invention differed from the earlier patents. A response was also required concerning the medical journal article listed in [56]. Because the patent eventually was approved, it is evident that the inventors managed to persuade the patent examiner that their idea was different from those described in the cited references.

The Abstract [57] is a short summary of what is described in the patent application.

On pages 1, 2, and 3 of Fig. 1-1, there are patent drawings. The parts of these drawings are labeled with numbers, which are referred to in the written description portion of the patent.

A detailed description of the invention, also known as the Specification, begins on page 4 of Fig. 1-1. The Specifi-

Fig. 2-1. Patent: Safety Control Valve for an Artificial Heart.

United States Patent [19]

Affeld et al.

[11] **Patent Number:** **4,559,648**

[45] **Date of Patent:** **Dec. 24, 1985**

[54] **SAFETY CONTROL VALVE FOR AN ARTIFICIAL HEART**

[75] Inventors: **Klaus Affeld; Axel Mohnhaupt; Frank Jorn,** all of Berlin, Fed. Rep. of Germany

[73] Assignee: **Thoratec Laboratories Corporation,** Berkeley, Calif.

[21] Appl. No.: **625,506**

[22] Filed: **Jun. 28, 1984**

[30]   **Foreign Application Priority Data**

Jun. 29, 1983 [DE]   Fed. Rep. of Germany 3323862

[51] Int. Cl.⁴ A61F 1/24
[52] U.S. Cl. 623/3; 91/28
[58] Field of Search .... 3/1.7; 128/1 D; 417/394; 91/28, 510

[56]   **References Cited**

U.S. PATENT DOCUMENTS

2,396,984 8/1944 Broadston et al. 91/28
4,173,796 11/1979 Jarvik 3/1.7

OTHER PUBLICATIONS

The International Journal of Artificial Organs/ vol. 5, No. 3, 1982/ pp. 157–159, "Completely Integrated Wearable TAH–Drive Unit", H. P. Heimes, F. Klasen.

Primary Examiner—Richard J. Apley
Assistant Examiner—James Prizant
Attorney, Agent, or Firm—Flehr, Hohbach, Test, Albritton & Herbert

[57]   **ABSTRACT**

A safety drive for an artificial heart, if the right drive should fail it is simply put out of operation, while if the left drive fails, the right drive, which is still functioning, is switched over to the left circulatory pump and thus serves as a system-internal backup drive.

7 Claims, 3 Drawing Figures

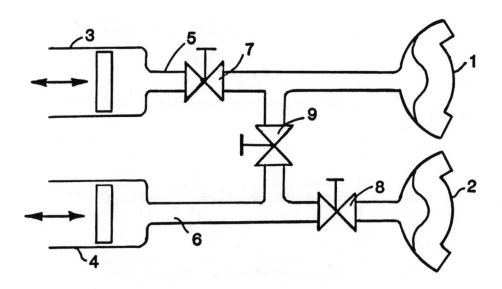

FIG. I

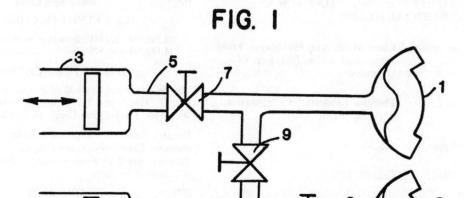

FIG. 2

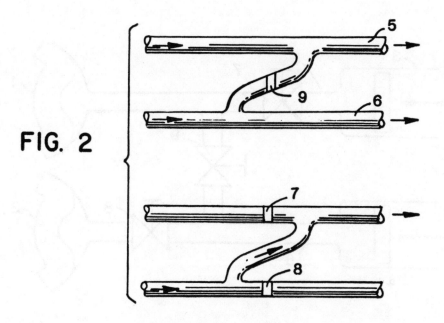

FIG. 3

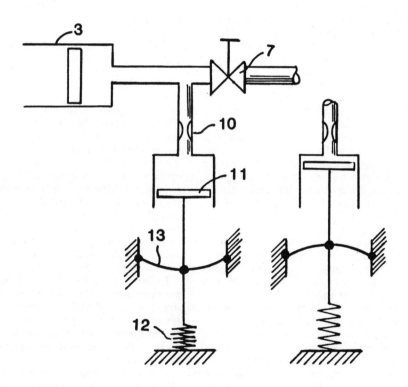

1

SAFETY CONTROL VALVE FOR AN ARTIFICIAL HEART

The invention concerns a drive with special safety reserves for a pneumatic artificial heart. Pneumatic artificial hearts are usually powered by a drive located outside the body. This drive must be distinguished by especially high operating reliability. In contrast to the case with other artificial organs, such as the artificial kidney, for example, any malfunctioning of the drive for an artificial heart carries with it an immediate risk of death. Besides having components specially selected to reliability, the drive must also have a structure which increases the safety of the system.

Drives for pneumatically driven circulatory pumps are available in which compressed air is modulated by pulse-actuated magnetic valves in such a way that the drives are able to power the pneumatic circulatory pumps. Another existing arrangement is to have this pulsed compressed air generated by reversing pneumatic pistons. In a further variant, reversing rotary compressors are used to generate the pulsed compressed air. Until now, increased safety could be achieved only by a very careful selection of components and by having a backup system available should one of the drives break down. These drives, however, are usually very cumbersome, so that it is barely possible for the patient to move about with even one drive, and with the backup system this becomes extremely difficult. Futhermore, when a drive fails two pneumatic lines must be replaced, entailing additional risk.

The present invention eliminates the problem of the possible lack of a backup system, while at the same time offering much higher system reliability than that of an individual drive. This problem is solved according to the invention in that individual drives are used for both the left and right ventricles; the drives are not coupled to each other either mechanically or electrically, but are interconnected by signal and air flow components in such a way that the drives for the left pump can be replaced by that for the right. This characteristic of the system is based on the practical finding that for a limited time the body is basically dependent only on the power of the pump to the left side and can do without the operation of the right pump for a more or less prolonged period of time. Thus if the right drive breaks down due to a mechanical or electrical failure there is no immediate danger to the patient, and there is ample time available to connect up a new drive. If the left drive fails, on the other hand, in the system according to the invention the right drive is switched over so that it serves as the drive for the left pump.

This switching can be done manually, after the user of the drive is alerted to a malfunction by an alarm, or it can be done automatically. The average pressure in the compressed-air tubing of the left drive can be used as an appropriate signal, this pressure clearly registering above zero when the left drive is functioning. If the left drive fails, then this pressure drops to a much lower value close to the zero point. This pressure drop can be used to make the switch over by means of an installed mechanism. The averaging of the pressure can be performed in a simple manner by a capillary and the switching itself by a spring component, which is held down by the average pressure in the left drive system during normal operation. The switch over should be executed so that the left pump is disconnected from the

2

right drive system and the defective left drive system is also disconnected from the lines to the left pump. It is useful to have the switch designed so that it functions bistably, that is, so that it is stable at the two end positions but unstable in the middle.

In another embodiment this switch over can also be accomplished by an electromechanical servo component, which does the switching based on an analysis of the pressure signals or of drive signals. After the switch over the right drive suddenly has to pump against a high pressure. This sudden assumption of the left pumping load can also be accomplished in an especially simple and reliable manner according to the invention if the drive is regulated to follow a reference volume curve rather than a reference pressure curve. In normal operation the two drives are synchronized with each other by having the left drive follow the left auricular pressure or the left pulmonary arterial pressure. These pressures can be measured either directly or indirectly, or they are determined by means of signals obtained from an analysis of the pneumatic drive pressure of the left pump in the diastolic phase.

The advantages which the invention is intended to offer consist particularly in the fact that the need to have available or carry along a complete backup system is eliminated, and nearly the same level of safety is achieved through a characteristic of the drive according to the invention.

Referring to the drawing:

FIG. 1 is a schematic diagram showing one embodiment of the invention.

FIG. 2 is a switching diagram showing the operation of the system.

FIG. 3 is a schematic diagram showing an automatic switch for use in accordance with the invention.

FIG. 1 shows the left circulatory pump 1 and the right pump 2, which are driven by a drive—here a piston—and 4. The pressure pulse is tansmitted by tubing lines 5 and 6. In normal operation valves 7 and 8 are open, while valve 9 is closed. Should the right drive 4 fail, the left drive 3 is able to take over the entire circulatory load. If the left drive 3 fails, however, it is disconnected from the circulatory pump 1 by the valve 7, and the right drive 4 is disconnected from the right pump 2 by the valve 8 and switched to the left pump 1 by the opening of the valve 9. In this way the right drive 4 takes over the entire circulatory load.

Referring to the switching diagram of FIG. 2, the top portion represents normal operation with lines 5 and 6 free-flowing and valve 9 closed to separate the two systems. The bottom portion shows the condition of the system if there is a malfunction: both pumps are closed off from their original drives by valves 7 and 8, while valve 9 is open.

Referring to FIG. 3, there is shown an automatic mechanical switch for use with the embodiment of FIG. 1. A branch with a capillary 10 is provided between left drive 3 and the valve 7. As long as drive 3 is able to function, a pulsed pressure prevails in the system, which is time-averaged by the capillary 10 and when at the time average exerts positive pressure on a piston 11 and thus compresses a spring 12. If drive 3 fails, the pulsed pressure is absent and the spring 12 presses the piston 11 back to the position shown at the right of FIG. 3. This motion of the piston is used to bring about the valve switching, as described above, in the manner according to the invention. Meanwhile a bistable spring element 13 ensures that the valve is able to assume only one of

4,559,648

<table>
<tr><td align="center">3</td><td align="center">4</td></tr>
</table>

the two positions—normal or disrupted—and prevents it from slipping into a middle position.

What is claimed is:

1. A safety drive for an artificial heart, comprising left and right pneumatic circulatory pumps each made of biocompatible material and each being driven by one of two mutually independent pneumatic drives, and means for switching the drive for said right pneumatic circulatory pump to serve as a backup drive for the left pneumatic circulatory pump.

2. A safety drive according to claim 1, wherein said means for switching is automatic and includes a mechanically bistable switch.

3. A safety drive according to claim 1 wherein said means for switching is manual.

4. A safety drive according to claim 2 wherein said switching means includes an isolated servo component responsive to an analysis of a compressed air signal from the left pneumatic circulatory pump.

5. A safety drive according to claim 2 wherein the left and right drives are regulated according to an air volume flow reference curve.

6. A safety drive according to claim 2 wherein, in normal operation, the volume flow of the left drive is regulated by the left auricular pressure, and upon failure of the left drive the right drive is also regulated by the left auricular pressure.

7. A safety drive according to claim 2 wherein the left air volume flow is regulated by a signal from an analysis of the pneumatic drive pressure of the left pump in the diastolic phase.

* * * * *

cation tells everyone how to make and use the invention. On the last page of Fig. 1-1 is the Claims section. Claims define exactly what is contained in the patent rights and what is not, like the legal description of a parcel of land in a deed.

It is obvious that a lot of time and effort was put into writing the abstract, specification, and claims and into making the patent drawings. Who must do all of this writing and drawing?

If you guessed the Patent Office does this for you, guess again. The text and drawings that appear in the issued patent are lifted from the patent application as it is submitted by the applicant. The applicant presents the Patent Office with text and drawings in the form of a patent application. The office reviews the application to determine if various requirements are met. If so, it issues a patent containing the text and drawings. If not, the applicant must amend the application until the requirements are met, or else appeal from the patent examiner's refusal to issue a patent.

CHAPTER 3

How to Do a Patent Search

In order to determine if your invention is new and therefore eligible for a patent, it is necessary to do a patent search.

Because any patent, printed publication, knowledge, or use can be prior art, a patent search, ideally, should cover all literature including such things as scientific journals and manufacturer catalogs. In theory, it should also require personal interviews with all engineers working in that field of technology, because prior "knowledge or use" need not be proven by a writing. In practice, a search is limited to (1) a computer search of on-line databases such as LEXPAT, DIALOG, CHEMICAL ABSTRACTS, etc. or (2) a manual search of patent indexes and patents.

COMPUTERIZED PATENT SEARCHING

Access to computerized searching facilities is available at most university and large public libraries. In general, you need only describe your invention to the reference librarian and pay the applicable search fee. The computer will produce a list of patents and scientific publications that contain particular key words. The librarian can locate the publications for you and, if need be, send away for copies of articles.

To inspect particular patents found by the computer, you can either (1) request copies of patents by number directly from the patent office or, (2) read summaries of the patents in a publication called the *Official Gazette*.

To request copies of patents, just send in a list of patent numbers and $1.50 per patent to the following address:

Commissioner of Patents and Trademarks
Washington, D.C. 20231

Delivery of patents usually takes about 3 weeks.

Reading summaries of patents in the *Official Gazette* is easier, faster, and cheaper. Most of the libraries listed in Fig. 3-1 have a complete set of *Gazettes* going back to the 19th century. In the *Gazettes*, summaries of each issued patent are arranged according to patent number. Included with most summaries is a diagram of the invention. A typical *Gazette* page is shown in Fig. 3-2.

CASSIS-ASSISTED PATENT SEARCHING AT LIBRARIES

A manual search for patents, with help from the free CASSIS database system, can be accomplished at any of the libraries listed in Fig. 3-1. This type of search involves the following steps.

1. Use the *Index of Classification* to find the general category of invention. Fig. 3-3 is a sample page from the *Index*. If, for example, your invention has to do with skis, simply refer to the heading "Ski" in the index and select the appropriate class and subclass numbers. Under the heading "Ski," several secondary classes such as "ski poles," "water skis" etc. are listed.

2. If your invention fits into one of the secondary classes listed in the index, skip directly to step 3. If not, then write down the main classification number and look in a more detailed index book, the *Manual of Patent Classification*.

   The main class for "Ski" listed in the *Index* is "280 601." Therefore, look in the *Manual* for Class 280 Subclass 601. The page from the *Manual* covering this class and subclass is shown in Fig. 3-4. As you can see, the *Manual* has a much more detailed listing of invention

STATE	NAME OF LIBRARY
Alabama	Auburn University Libraries
	Birmingham Public Library
Alaska	Anchorage Municipal Libraries
Arizona	Tempe: Science Library, Arizona State University
Arkansas	Little Rock: Arkansas State Library
California	Los Angeles Public Library
	Sacramento: California State Library
	San Diego Public Library
	Sunnyvale: Patent Information Clearinghouse
Colorado	Denver Public Library
Delaware	Newark: University of Delaware
Florida	Fort Lauderdale: Broward County Main Library
	Miami-Dade Public Library
Georgia	Atlanta: Price Gilbert Memorial Library, Georgia Institute of Technology
Idaho	Moscow: University of Idaho Library
Illinois	Chicago Public Library
	Springfield: Illinois State Library
Indiana	Indianapolis: Marion County Public Library
Louisiana	Baton Rouge: Troy H. Middleton Library, Lousiana State University
Maryland	College Park: Engineering and Physical Sciences Library, University of Maryland
Massachusetts	Amherst: Physical Sciences Library, University of Massachusetts
	Boston Public Library
Michigan	Ann Arbor: Engineering Transportation Library, University of Michigan
	Detroit Public Library
Minnesota	Minneapolis Public Library & Information Center
Missouri	Kansas City: Linda Hall Library
	St. Louis Public Library
Montana	Butte: Montana College of Mineral Science and Technology Library
Nebraska	Lincoln: University of Nebraska-Lincoln, Engineering Library
Nevada	Reno: University of Nevada Library
New Hampshire	Durham: University of New Hampshire Library
New Jersey	Newark Public Library
New Mexico	Albuquerque: University of New Mexico Library
New York	Albany: New York State Library
	Buffalo and Erie County Library
	New York Public Library (The Research Libraries)
North Carolina	Raleigh: D. H. Hill Library, N.C. State University
Ohio	Cincinnati & Hamilton County, Public Library of Cleveland
	Columbus: Ohio State University Libraries
	Toledo/Lucas County Public Library
Oklahoma	Stillwater: Oklahoma State University Library
Oregon	Salem: Oregon State Library
Pennsylvania	Cambridge Springs: Alliance College Library
	Philadelphia: Franklin Institute Library
	Pittsburgh: Carnegie Library of Pittsburgh
	University Park: Pattee Library, Pennsylvania State University
Rhode Island	Providence Public Library
South Carolina	Charleston: Medical University of South Carolina
Tennessee	Memphis & Shelby County Public Library and Information Center
	Nashville Vanderbilt University Library
Texas	Austin: McKinney Engineering Library, University of Texas
	College Station: Sterling C. Evans Library, Texas A & M University
	Dallas Public Library
	Houston: The Fondren Library, Rice University
Utah	Salt Lake City: Marriott Library University of Utah
Washington	Seattle: Engineering Library, University of Washington
Wisconsin	Madison: Kurt F. Wendt Engineering Library, University of Wisconsin
	Milwaukee Public Library

Fig. 3-1. List of libraries at which patent searches can be performed.

4,643,306
POSTAL TRAY
Patrick Ryan, Reseda, Calif., assignor to Alpha Mail Systems,
Newport Beach, Calif.
Filed Nov. 8, 1985, Ser. No. 796,248
Int. Cl.⁴ B65D *1/34, 21/02*
U.S. Cl. 206—425   **5 Claims**

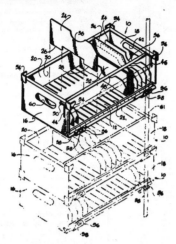

1. A postal tray facilitating the sorting and distribution of
mail comprising:
a housing, said housing having a planar bottom wall which
has a peripheral edge, a pair of planar side walls attached
to said peripheral edge and extending from said bottom
wall, a planar back wall attached to said peripheral edge
and extending from said bottom wall, said back wall located between said side walls, said housing having an open
top and open front, an internal chamber defined by said
bottom wall and enclosed by said side walls and said back
wall;
an opening assembly formed within said bottom wall, said
opening assembly connecting with said internal chamber;
a plurality of planar partition members removably mounted
within said opening assembly, said partition members
extending within said internal chamber, pockets being
formed between directly adjacent partition members, said
pockets being adapted to receive postal envelopes; and
a rod connected to said side walls and extending therebe-
tween, said rod being located directly adjacent said top
and directly adjacent said front, said rod being movable
with respect to said side walls between an upper position
and a lower position, said upper position locating said rod
across said top, said lower position locating said rod
across said front.

4,643,307
**PACKING ARRANGEMENT FOR ARTICLES OF
DIFFERENT SIZE**
Don Wilkinson, 32 S. Palm Ave., Sarasota, Fla. 33577
Filed Feb. 7, 1986, Ser. No. 827,071
Int. Cl.⁴ B65D *85/20, 85/62;* B65B *59/00*
U.S. Cl. 206—443   **12 Claims**
1. A packing arrangement comprising:
a plurality of bodies having circular cross-sections;

said circular cross-sections having radii proportional to one,
two and three; and

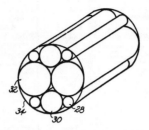

each of said bodies having one of said radii being tangent to
at least two of said bodies having the other two radii.

4,643,308
SHIELDED DISKETTE CASE
Thomas J. Michel, Miami, Fla., assignor to Data Medi-Card,
Inc., Lake Worth, Fla.
Filed Feb. 5, 1986, Ser. No. 826,309
Int. Cl.⁴ B65D *85/57, 47/24*
U.S. Cl. 206—444   **6 Claims**

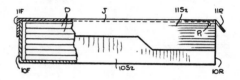

1. A case for protectively housing diskettes having data
magnetically recorded thereon, the case also acting to shield
the diskettes from stray magnetic fields, electrostatic discharges and ionizing radiation, said case comprising:
A a box adapted to house the diskettes; and
B a cover hinged to the rear of the box, the cover being
provided with front and side flanges which, when the
cover is closed, lie against the corresponding walls of the
box, and with an outwardly inclined rear baffle which,
when the cover is open, lies against the rear wall of the
box, and when the cover is closed is angled with respect to
said rear wall to permit the cover to swing open, said box
and said cover including said flanges and said baffle being
formed of a paramagnetic metallic material which is not
substantially penetrated by said stray magnetic fields,
electrostatic discharges and ionizing radiation, the angle
assumed by the baffle when the cover is closed being such
as to prevent penetration into said box, whereby the diskettes housed in the box are fully shielded.

4,643,309
FILLED UNIT DOSE CONTAINER
Hans C. A. Evers, Södertälje, Sweden, assignor to Astra
Lakemedel AB, Sweden
Continuation of Ser. No. 713,999, Mar. 20, 1985, abandoned, and
a continuation-in-part of Ser. No. 461,594, Jan. 27, 1983. This
application Mar. 21, 1986, Ser. No. 844,215
Claims priority, application Sweden, Feb. 8, 1982, 8200720
Int. Cl.⁴ B65D *73/00*
U.S. Cl. 206—484   **10 Claims**
1. A filled thermoplastic solution container containing a
single dose of a sterile medication for administration to a
human or animal patient by injection using a hypodermic
syringe comprising a syringe having a conical tip to which a
hypodermic needle is attached, said container having a neck
portion with an inner surface which has converging and diverging sections which intersect and define an orifice at their
intersection, the diverging section extending outwardly from

Fig. 3-2. Typical page from Official Gazette.

	Class	Subclass
Vehicle	116	28R+
Watches alarm	368	244
Water heater	126	388
Water heater vessel combined	126	388
Signaling Teaching	434	223
Signature Gatherers and Conveyers		
Conveyor power driven	198	644
Machine for writing several	33	23.6
Sheet associating	270	54+
Signing machine	D18	21
Silastic T M (See Synthetic Resin or Natural Rubber)		
Silazanes	568	410+
Silencer (See Insulating, Sound; Muffle; Noise Deadening)		
Firearm and gun	89	14.4
Muffler for firearm	181	223
Silent Butler	D32	74
Silica	423	335
Abrasive compositions containing	51	308
Absorbents containing	502	407+
Aerogel	106	287.34
Catalysts containing	502	232+
Cement containing	106	98
Gel	423	338
Dispersing or stabilizing agent combined	252	315.1
Glass compositions containing	501	53+
Glass manufacturing	65	
Lime compositions containing	106	120
Lubricants containing	252	9+
Silicane	556	465+
Silicates & Silicon Compounds,		
Inorganic	423	324+
Abrasive compositions containing	51	308
Absorbents containing	502	407+
Catalysts containing	502	232-
Coating or plastic compositions containing		
Alkali metal	106	74+
Carbide	501	87+
Glass	501	11+
Lime combined	106	120
Opacifiers for enamels	106	312
Pigments fillers aggregates	106	291
Portland cement type	106	89+
Portland cement type making	106	100+
Refractories	501	94+
Slag	106	117
Detergents containing	252	89.1+
Dyeing	8	523+
Fertilizers containing	71	
Hydrofluosilicic acid	423	341
Lubricants containing	252	9+
Organic compounds	556	400+
Esters	556	482-
Silica	423	335
Silicofluorides	423	324
Silicic Acid	423	325
Silicofluoric Acid	423	341
Silicofluorides	423	324+
Silicol	556	463
Silicon	423	348
Alloys		
Aluminum	420	548
Aluminum copper	420	537
Copper	420	490
Metallic	420	578
Carbide	423	345
Ferro producing	420	578
Misc. chemical manufacture	156	DIG. 64
Organic compounds in compositions	106	287.1+
Polymers (See synthetic resin or natural rubber)		
Room-temperature curable	528	901*
Silicon Steel	148	110+
Silicone	556	465+
Compositions	106	287.13
Sililation of Polymers	525	100+
Silk		
Artificial making		
Apparatus	425	66
Apparatus	425	67+
Apparatus	425	76
Apparatus	425	382.2
Processes	264	165+
Chemical treatment	8	128R+
Degumming	8	138
Fabrics woven	139	428
Fermentative treatment	435	268
Liberating	19	3

	Class	Subclass
Screen printing	101	114+
Process	101	129
Thread finishing	28	217+
Silkworm Culture	119	6
Sill	52	204
Railway car	105	396+
Weatherstrips	49	303+
Sillimanite		
Ceramic containing	501	141+
Porcelain containing	501	141+
Refractory containing	501	128+
Silos		
Circular	52	245+
Compacting ensilage	100	65+
Portable, & means for erecting	414	919*
Prestressed	52	224
With port	52	192+
Siloxanes	556	450+
Resinous solids (See also synthetic resin or natural rubber)	528	10+
Silver	75	
Alloys	420	501
Chest or box for	D 3	75
Composite metal stock	428	673
Electrodeposition	204	46.1
Electrolytic synthesis	204	109+
Mercury alloy	204	126+
Hydrometallurgy	75	118R
Misc. chemical manufacture	156	DIG. 101
Organic compounds	556	110+
Proteins	530	400
Pyrometallurgy	75	83
Silverware		
Chest or box for	D 3	75
Sorter	209	926*
Tray or container	206	557
Simulations		
Article dispensing	221	24
Cabinets with	312	204
Display	40	538+
Material dispensing combined	222	80+
Pyrotechnics	102	335
Supports digest	248	DIG. 8
Toys	446	
Velocipedes	280	1.1 R+
Sine Bar	33	536+
Sinew Removing	17	11.3
Singeing		
Cloth finishing	26	3-
Electric burner	219	223+
Hair combs for	132	118
Hogs	17	20
Thread finishing	28	239
Warps	28	174
Single Grid Twin Plate Control Tube	313	303
Single Roll Crusher	241	221+
Singletree		
Horse detacher		
From thill or evener	278	32
Traces from singletree	278	24+
Sink (See Receptacles)	4	
Cabinet combined	312	228
Design	D23	58+
Disappearing bowl	4	DIG. 2
Drain board or tray	D32	56
Lining mat	D32	57
Strainers and stoppers	4	286+
Structure	4	619+
Traps	4	191+
Tub combined	4	553+
Sinker		
Fishing type	43	43.1+
Design	D22	145
Knitting machine	66	104+
Traps	43	68
Sinkhead	249	106
Sintering		
Apparatus, ore	266	176
Stationary grate	266	185
Beneficiating ores by	75	5
Cellular synthetic resin or natural rubber product	521	919*
Ceramic composition	501	1
Clay	264	56+
Glass granules	65	18.1
Apparatus	65	144
Portland type cement making	106	100+
Powder metallurgy		
Processes	419	
Liquid phase	419	47
Products	75	228+

	Class	Subclass
Processes	432	1+
Lime gypsum cement	106	100+
Sinusoidal Generator		
Electronic tube type oscillator	331	
Generator structure	310	
Generator system	322	
Siphon		
Acetylene generator	48	17
Boiler	122	68
Bottle	215	4+
Design	D 7	51
Filling	141	14+
Bowl	4	421+
Code recorder	178	91
Dispensers	222	416
Discharge assistant combined	222	204
Fluid flow meter	73	222
Fluid handling	137	123+
Automatic	137	132+
Lubricator	184	85
Plumbing	137	247.11+
Syringes	604	131+
Water closet		
Siphon bowl	4	421-
Tank	4	368+
Siren	116	147
Design	D10	120
Electric	340	405
Toy	446	205
Sirup (See Syrups)		
Sizing (See Assorter)		
Bread, pastry and confections	99	352+
Chain making	59	29
Coating compositions		
Natural resin containing	106	238
Protein containing	106	124+
Starch containing	106	210+
Coating processes	427	
Dyeing processes combined	8	495+
Finger ring forming and	29	8
Hat making	223	10
Paper tube making	493	269
Textiles chemical modification of	8	115.6
Warp threads	28	178
Skate and Skaters Appliances	280	11.1 R
Clamp for sharpening by abrading	51	228
Design	D21	224+
Ice, powered	180	180+
Roller, powered	180	180+
Sharpening by cutting	76	83
Skaters supports	272	70+
Snow	280	600
Design	D21	224+
Strap	280	11.3
Water	441	65+
Wheels for	301	5.3+
Skeeball Game	273	352
Skein		
Axles	301	134+
Fluid treating package	8	155.2
Reeling	441	26+
Holders	242	127
Skeining	28	291
Skeleton		
Figure toys	446	373+
Keys	70	394
Plug valve	251	310+
Towers	52	648+
Skelping Metal	72	176+
Skewer Pins	17	1S
Ski (See Aquaplane; Sled)	280	601+
Apparel design	D 2	31
Bindings	280	611+
Clamp type having plural diverse axes clampsps	280	DIG. 12
Controlled by skier postion or by remote control	280	DIG. 13
Brake	280	604+
Carriers	224	917*
Carriers, hand	294	147
Case	206	315.1
Clamp	280	814
Design	D21	229+
Handle or grip design	D 8	DIG. 6
Poles and sticks	280	819+
Presses	280	815
Snow, ie, non-water, powered	180	180+
Shoe fasteners	280	611+
Trail groomer	37	219+
Teaching skiing	434	253
Water skis	441	68+

Fig. 3-3. Page from Index of Patent Classification.

```
1         MISCELLANEOUS
1.1 R     SIMULATIONS, I.E., VELOCIPEDE TYPE
1.1 A     ..Figure with hand towed, attendant
             stabilized chair
1.11 R    .Occupant propelled
1.11 A    ...Boat simulating
1.12      ..Aircraft
1.13      ..Figure
1.14      ...Noise makers
1.16      ...Figure and carriage
1.165     ....With moving figure or figure parts
1.167     .....Relatively movable legs
1.173     ...Plural
1.175     ...Progressing rocker type
1.177     ....Plural relatively moving rockers
1.181     ...With alternately advanced ground
             engaging means
1.182     ....Galloping progression
1.183     .....Foot or saddle actuated
1.184     .....With additional ground engaging
             means
1.186     ...Occupant not astride
1.188     ...Convertible, interchangeable or
             adjustable
1.189     ...Combined
1.191     ...With rider body motion
1.192     ....With figure motion
1.193     .....Platform rocker type
1.194     .....Additional relatively movable drive
1.195     .....Rigid or rigid link steering
1.196     .....Body centrally pivoted
1.201     ...Movably mounted figure or figure
             parts
1.202     ....Steering movement
1.203     .....Body hinged in vertical plane
1.204     ....Movable legs
1.206     ...Rein steered or driven
1.207     ...Stirrup steered or driven
1.208     ...Hollow body
1.21      .Aircraft
1.22      .Figure
1.5       WITH MEANS FOR ENGAGING BODY OF WALKING
             ATTENDANT
3         WHEELED PROPELLED
4         HOSE AND LADDER
5 R       TANK OR BOILER
5 F       ..Tank forms load support
5 G       ..Insulated tank walls
5 H       ..Saddle type; vehicle frame straddling
             type
5 A       ..Automobile and vehicle service tank
             type
5 B       ..Ground engaging and rolling wheel
             cylinder type
5 C       ..Semitrailer tank vehicle
5 D       ..Truck with tank (i.e., tank truck)
5 E       ..Trailer tank vehicle (other than
             semi-trailer type)
7         .With body leveling devices
5.2       STEP OR ABUTMENT ASCENDING TYPE VEHICLE
5.22      .With endless track
5.24      .With skid or rollerway
5.26      .With spider-type engaging means
5.28      .Relatively movable running gear
             portions
5.3       .Jack-type attachment, i.e., elevating
5.32      .Fulcrum attachment
6 R       BODY-LEVELING DEVICES
6 H       ..Vehicle leveling relative to
             horizontal by hydraulic motor
6.1       .With ground engaging means moved
             vertically responsive to body level
             sensing means
6.11      .With ground engaging means
             interconnected for opposite vertical
             adjustment
7.1       CONVERTIBLE, I.E., VELOCIPEDE TYPE
7.11      .To drop frame
```

```
7.12      .Wheel to or from runner
7.13      ..Skates
7.14      ..Substituted
7.15      .Occupant propelled
7.16      ..To or from plural occupant
7.17      ..To non-propelled
8         WHEELED AND RUNNER
9         .Retracting wheel or runner
10        ..Wheel-runner type
11        ..Pivoted lateral axis
11.1 R    SKATES
11.1 BR   ..Ball-type rollers
11.1 BT   ..Roller skis
11.1 ET   ..Endless tread type
11.115    .With propulsion means driven by
             occupant
11.12     .Runner type
600       ..Snow skates
601       ..Skis
602       ...With camber or flexibility control
603       ...Knockdown, folding or collapsible
604       ...With climbing or braking means
605       ....Pivotally mounted brake member
606       ...Steerable or with handle
607       ...With foot supporting plate or portion
608       ...With edge or guide strip
609       ...With special shape, contour or groove
610       ...Laminated or synthetic material
11.14     ..With resiliently mounted foot supports
11.15     ..Jointed runner and foot supports
11.16     ..Extensible
11.17     ..Tubular foot posts
11.18     ..Runners
11.19     .Wheeled
11.2      ..With brake
11.21     ..With retrogression preventers
11.22     ..Tandem wheels
11.23     ..Two-wheeled
11.24     ..One-wheel
11.25     ..Two-wheeled
11.26     ..Extensible
11.27     ..Trucks and mountings
11.28     ...Resiliently yieldable
11.3      .Shoe attaching means
11.31     ..Toe and/or heel clamps
11.32     ...Screw actuated
11.33     ...Lever actuated
11.34     ....With screw adjustment
611       ..Ski fasteners
612       ...Magnetic
613       ...Operating mechanism located in or
             under boot
614       ...Cross-country to or from downhill
615       ...Cross-country only
616       ...Toe and heel fasteners interconnected
             for simultaneous operation
617       ...Toe and heel fasteners mounted on
             common support plate or element
618       ....Plate is movable or releasable from
             ski
619       ...Heel cable and/or ankle strap type
620       ....Attached to movable or releasable
             plate
621       ....Cable tightening mechanism mounted
             on ski
622       ....With means to separate cable or pull
             cable from boot
623       ...Hold-downs or clamps
624       ....Side or rear hold-down only, e.g.,
             "Spademan" bindings
625       ....Clamp comprises plural sole engaging
             members
626       ....Pivotal about horizontal and/or
             vertical axis
627       .....With non-pivotal hold-down or clamp
628       .....Pivotable about both horizontal and
             vertical axis
```

Fig. 3-4. Page from Manual of Classification.

```
CLSF SELECTED: 280/610
NUMBER OF PATENTS: 213
     1.  4,655,473  OR              6.  4,634,563  XR
     2.  4,652,006  XR              7.  4,634,140  OR
     3.  4,647,063  OR              8.  4,627,635  XR
     4.  4,639,009  XR              9.  4,595,215  XR
     5.  4,635,954  XR             10.  4,556,237  OR
CLASSIFICATION MODE-L11-80
    11.  4,545,597  OR             46.  4,140,330  OR
    12.  4,540,195  OR             47.  4,135,732  OR
    13.  4,537,417  XR             48.  4,118,051  OR
    14.  4,530,871  XR             49.  4,118,050  XR
    15.  4,523,772  OR             50.  4,094,528  OR
    16.  4,518,453  XR             51.  4,093,268  OR
    17.  4,498,686  OR             52.  4,077,652  OR
    18.  4,455,037  OR             53.  4,071,264  OR
    19.  4,440,418  XR             54.  4,068,861  OR
    20.  4,438,946  XR             55.  4,065,150  OR
    21.  4,420,523  XR             56.  4,047,735  OR
    22.  4,416,929  XR             57.  4,044,083  XR
    23.  4,412,687  OR             58.  4,035,000  OR
    24.  4,409,287  XR             59.  4,026,575  OR
    25.  4,405,149  XR             60.  4,007,946  XR
    26.  4,386,982  XR             61.  4,005,875  OR
    27.  4,383,701  OR             62.  3,997,178  OR
    28.  4,382,610  OR             63.  3,980,312  OR
    29.  4,349,212  OR             64.  3,970,324  OR
    30.  4,337,963  XR             65.  3,967,992  XR
    31.  4,314,384  XR             66.  3,958,810  OR
    32.  4,313,614  OR             67.  3,949,988  XR
    33.  4,293,142  OR             68.  3,947,049  XR
    34.  4,272,578  XR             69.  3,944,239  XR
    35.  4,272,577  XR             70.  3,940,157  OR
    36.  4,270,768  OR             71.  3,933,362  OR
    37.  4,261,778  XR             72.  3,930,658  OR
    38.  4,260,576  XR             73.  3,928,106  XR
    39.  4,259,274  XR             74.  3,918,731  OR
    40.  4,233,098  XR             75.  3,918,728  XR
    41.  4,209,867  XR             76.  3,915,465  XR
    42.  4,175,767  OR             77.  3,902,732  OR
    43.  4,165,886  XR             78.  3,901,522  OR
    44.  4,154,459  OR             79.  3,897,074  OR
    45.  4,146,251  OR             80.  3,894,745  OR
CLASSIFICATION MODE-L81-125
    81.  3,893,681  OR            104.  3,707,296  OR
    82.  3,879,245  XR            105.  3,705,729  OR
    83.  3,861,699  OR            106.  3,704,023  OR
    84.  3,844,576  OR            107.  3,698,731  OR
    85.  3,832,251  XR            108.  3,652,102  OR
    86.  3,823,956  OR            109.  3,635,484  OR
    87.  3,816,573  XR            110.  3,635,483  OR
    88.  3,814,452  OR            111.  3,635,482  OR
    89.  3,807,746  OR            112.  3,628,802  OR
    90.  3,806,142  OR            113.  3,614,116  OR
    91.  3,801,116  OR            114.  3,612,556  OR
    92.  3,790,184  OR            115.  3,567,237  OR
    93.  3,776,563  XR            116.  3,549,461  XR
    94.  3,774,254  XR            117.  3,542,388  OR
    95.  3,771,805  OR            118.  3,503,621  OR
    96.  3,762,734  OR            119.  3,501,161  OR
    97.  3,740,301  XR            120.  3,498,626  OR
    98.  3,738,675  OR            121.  3,493,240  OR
    99.  3,736,609  XR            122.  3,475,035  OR
   100.  3,734,519  OR            123.  3,457,129  XR
   101.  3,733,380  XR            124.  3,438,828  XR
   102.  3,727,936  OR            125.  3,416,810  OR
   103.  3,722,901  OR
CLASSIFICATION MODE-Q
```

Fig. 3-5. CASSIS search printout.

subclasses. Find the most relevant subclass number and write it down.

For example, if your invention is a teflon-coated snow ski, inspection of Fig. 3-4 should lead you to subclass 610, which covers "... laminated or synthetic material."

3. Once you have the proper class and subclass numbers, ask the reference librarian to do a CASSIS computer printout of all patents in that classification. CASSIS terminals are available at all libraries listed in Fig. 3-1 and can be used for free. A CASSIS printout for snow skis is shown in Fig. 3-5. You now have a complete list of all patents in a particular category.

4. Using your CASSIS list of patents, consult the *Official Gazette* or send in for copies of patents as is described in the previous section regarding computerized patent searches.

Notice that some patents' numbers listed in the CASSIS printout are listed as "OR" and others as "XR." The "OR" patents are those directly in the subclass you requested. The "XR" patents are cross references to related subclasses. If you find from your inspection of the *Gazettes* that the "OR" patents are not exactly what you want, try looking up the "XR" patents. If you find an "XR" that is more appropriate, write down the class and subclass of this patent and request a second CASSIS search.

SEARCHING AT THE PATENT OFFICE

Performing a search at the patent office public search room, located in Crystal Plaza, Arlington, Va., is exactly like doing a manual search as described above except you have available actual copies of all issued patents. After you find the proper class and subclass in the *Index* and *Manual*, you can go directly to a box containing all patents in this classification and read to your heart's content. The tedious step of getting a CASSIS printout and sifting through numerous volumes of the *Gazette* is eliminated.

But you can eliminate even the step of searching through the *Index* and *Manual*. Available in the public search room are several computer terminals that allow you to key-word search the *Manual of Classification*. By simply putting in key words describing your invention, you can get the patent office computer to find the proper class and subclass for you. This eliminates most of the difficult work.

If you cannot find the proper class and subclass for your search by means of the Index or computer terminals, find a patent examiner who works in the relevant field of technology. Examiners are usually willing to provide patent searchers with helpful hints. Often, if you tell the examiner about your idea, he or she will lead you directly to the most relevant prior art. You can find examiners in various fields of technology by consulting directories available in the patent office public search room. The examiners are located in adjacent buildings.

CHAPTER 4

Getting a Patent: A Four-Step Procedure

Getting a patent involves four steps:

Step 1. Write a document called a patent application, This contains a complete description of your invention and is ordinarily accompanied by rough drawings.

Step 2. Send the patent and accompanying papers to the Patent Office in Washington, D.C. The package to be filed includes the following:
- ✍ Patent Application
- ✍ Rough Drawings
- ✍ Abstract of the Disclosure
- ✍ Declaration for a Patent Application
- ✍ Verification of Small Entity Status
- ✍ Filing Fee
- ✍ Patent Application Transmittal Letter

Step 3. Respond to the patent examiner if he rejects or objects to portions of your application, or makes other requirements.

Step 4. Respond to the Notice of Allowance of your application by making any final changes that are required, paying an issue fee, and submitting final drawings. Upon receipt of the issue fee, the Patent Office sends the original copy of the patent to you.

REQUIREMENTS FOR A PATENT APPLICATION

Details regarding how to accomplish the four steps listed above will be given in the following chapters. To better understand these details, you should first understand the fundamental requirements that must be met by a patent application. The requirements are as follows:

- ❢ The invention must fit into one of the following categories: (1) process; (2) machine; (3) manufacture; (4) composition of matter; or (5) improvement of a previously existing process, machine, manufacture or composition of matter.
- ❢ The invention must have utility.
- ❢ The invention must be new.
- ❢ The invention must be more than an obvious variation of a previously existing invention.
- ❢ The patent application must clearly describe how to make and use the invention.
- ❢ The patent application must clearly explain what the inventor thinks his invention is. It must define exactly what it is about the invention that is different from what has been invented before.

The following discussion can help you decide whether your invention is eligible for a patent and also give you information regarding the application process.

Requirement A: Categorization

The invention must fit into one of the following categories: (1) process; (2) machine; (3) manufacture; (4) composition of matter; or (5) improvement of a previously existing process, machine, manufacture, or composition of matter.

WHAT IS A PROCESS?

According to a definition in the patent law, a process is "a process, art or method, and includes a new use of a known process, machine, manufacture, composition of matter, or material."

A *process* or *method* is typically described by listing steps for doing something. For example, a method for making boiled eggs includes the steps of (1) heating water to boiling; (2) placing the eggs in the water; and (3) removing the eggs after 3 minutes. If a method accomplishes something new, it might be patentable.

A process can also be a new use for an existing process, machine, manufacture, composition of matter, or material. For example, it is well known that compound X is a good material for use in laundry detergent. However, you have discovered that compound X is lethal to corn borers and would make a good insecticide. Therefore, you can get a patent on the use of compound X as an insecticide. Your new-use patent will not prevent others from making or selling compound X. However, it will prevent them from marketing, packaging, or preparing it as a pesticide.

To be eligible for a patent as a process or method, an invention must be a part of useful technology, as opposed to something that is not technology, such as business management or pure mathematics. If your idea is a mere method of doing business, it will be rejected by the Patent Office as "non-statutory subject matter" because it does not fall within the definition of patentable matter contained in the patent statute. For example, a method of operating a pizza parlor wherein you offer free home delivery would be a mere method of doing business. This idea, although a method, does not fit within one of the five categories of patentable inventions, as intended in the patent law. Likewise, a pure mathematical process, such as a method for solving a differential equation, will be rejected as non-statutory subject matter. The same goes for a pure scientific theory, such as $E = mc^2$.

For an idea to be a patentable process, it must have some impact upon technology. Perhaps the pizzas of the previous example must be boxed in a certain way to keep them warm. Thus, you have a new method for boxing pizzas, which is patentable. Or, your new method for solving a differential equation means that a computer can now control a wind tunnel blower more effectively. Thus, you have a new method for controlling a wind tunnel, which is patentable. Or, you have found out how the equation $E = mc^2$ applies to the construction of a reactor that makes electricity. Such a reactor could be patentable as a machine and the method by which it makes electricity could be a patentable process.

Lately, the Patent Office has been granting patents for useful mathematical algorithms, provided that these are described as methods. For example, the Patent Office recently granted a patent for a fast alternative to the Fourier transform, as is seen in Fig. 4-1.

WHAT IS A MACHINE OR MANUFACTURE?

The words *machine* or *manufacture* in patent law usage mean about the same as in ordinary usage. The word machine calls to mind a device that moves or transforms forces, such as a mechanical watch. A manufacture calls to mind a physical object that does not necessarily have moving parts, such as an ash tray or transistor. The Wright brothers flying machine, Patent No. 821,393, is a typical machine. Edison's light bulb, Fig. 4-2, is a typical manufacture.

A key point regarding all inventions, but particularly machines and manufactures, is that they must be created by man, rather than by nature. Suppose that you discover a species of bacteria under some rock that makes a new kind of antibiotic. Can you get a patent on this bacteria? The answer is no. The bacteria were invented by nature, not by you. Since you are the *discover*, not the *inventor*, you can't get a patent.

If you want some sort of patent, you must add value to that which was given to mankind by nature. Although you can't patent the bacterium as a machine or manufacture, you can apply for a patent on the process of growing this particular bacterium in vats. Since the bacteria in nature did nothing more than output their antibiotic into the local soil, you might also be able to get a patent on the purified antibiotic compound, *and* on a new use for the natural compound as an antibiotic. You have added value by (1) inventing a way to produce the compound commercially; (2) making a previously unknown substance, i.e., purified antibiotic or (3) finding a new use for that which was created by nature. You can also patent a bacterium with artificially altered DNA. Recently, patents have been authorized for genetically altered higher animals.

If your invention is mere printed matter, it will also be rejected as not being in a patentable category (i.e., as non-statutory subject matter). The mere arrangement of printed matter, although seemingly a "manufacture," is not considered to be within one of the classes. You are limited to copyright protection for this. However, if your printed matter is something more than a mere arrangement of print, you could have a qualified manufacture. Checks with perforated tear-off stubs, carbonless copies, or forger-proof paper fall within the class of patentable manufactures. Labels for grocery items printed with machine-readable bar codes would also qualify. In each of these examples, the invention has something more to it than mere printed words on paper. The game Monopoly would appear to be nothing more than a printed game board and printed cards. However, this was also not considered to be mere printed matter, as the issued patent, Patent No. 2,026,082, illustrates. Perhaps the little plastic hotels made the difference!

If the printed matter folds up into a useful object, such as the McDonald's extra large fries box, (seen in Fig. 4-3), it can also be patented.

WHAT IS A COMPOSITION OF MATTER?

An invention that is a man-made chemical compound, for example, acetylsalicylic acid (aspirin), qualifies as a patentable composition of matter. Likewise, if you use a nuclear accelerator to create a new isotope of carbon or your steam press to create a new type of potato chip, you will also have an invention in the composition of matter category. Physical mixtures of numerous compounds, as in cough syrup or gun powder; mixtures of metals as in alloys; and substances such as polymers, ceramics, and glasses also qualify.

(a) signals representative of the status of the aircraft's control surfaces,
(b) a command airspeed value; and
(c) the command input means produced command signals; memory means for storing the equations of flight of the aircraft under test, and logic means responsive to said input means and said memory means for producing said simulated pitch, roll and azimuth signals.

4,646,256
COMPUTER AND METHOD FOR THE DISCRETE BRACEWELL TRANSFORM
Ronald N. Bracewell, Stanford, Calif., assignor to The Board of Trustees of the Leland Stanford Junior University, Palo Alto, Calif.

Filed Mar. 19, 1984, Ser. No. 590,885
Int. Cl.⁴ G06F 15/31
U.S. Cl. 364—725   43 Claims

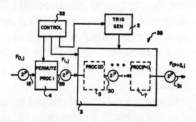

1. A method of performing a discrete transform on an input sequence of N input data values, $F_{0,j}$, where N is equal to 2^P, where P is a positive integer and where j has values from 0 to $N-1$, comprising,
permuting in permuting means the input sequence, $F_{0,j}$, with a bisecting permutation to form a permuted sequence, $F_{1,j}$,
processing in processing means said permuted sequence in P subsequent stages s where s has values from 2 to $P+1$, where the outputs from one stage form the inputs for the next stage and where each stage calculates N values of $F_{s,j}$ as a function of direct values from the previous $s-1$ stage, as a function of values from the previous $s-1$ stage multiplied by cosine factors to form cosine terms, and as a function of values from the previous $s-1$ stage multiplied by sine factors to form sine terms whereby the transformed data values, $F_{P+1,j}$, are provided from the $P+1$ stage.

4,646,257
DIGITAL MULTIPLICATION CIRCUIT FOR USE IN A MICROPROCESSOR
Daniel L. Essig; Luat Q. Pham; Joe F. Sexton, all of Houston, Tex., and Graham S. Tubbs, Tempe, Ariz., assignors to Texas Instruments Incorporated, Dallas, Tex.

Filed Oct. 3, 1983, Ser. No. 538,634
Int. Cl.⁴ G06F 7/52
U.S. Cl. 364—760   14 Claims

1. A digital multiplication circuit comprising:
a Booth recoder means for recoding a multiplier into N Booth operation sets where N is a positive integer that equals one half the number of bits in the multiplier;
a plurality of N partial product selector means having inputs and outputs connected in cascade arrangement from output to input with the Nth output being an intermediate output of the digital multiplication circuit, the cascade arrangement being N multiplicand sets of M bits in length and each member of the plurality of N partial product selector means being connected to a member of the N operation sets for implementing the recoded Booth operation set on a multiplicand set and where M is a positive integer;
summation means for summing the contents of the plurality of N partial product selector means the summation means

including a second plurality of summing means with predetermined members of the second plurality of summing means being located between predetermined members of the plurality of N partial product selector means; and
a domino means, operatively connected to the plurality of partial product selector means, for generating a plurality of evaluation pulses with each single evaluation pulse of

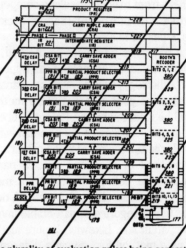

the plurality of evaluation pulses being generated to represented a worse case signal propagation time delay through a predetermined partial product selector means and connected to evaluate the predetermined member of the plurality of partial product selector means to provide outputs therefrom of each single partial product selector means when evaluated.

4,646,258
SAMPLED N-PATH FILTER
Saul Miodownik, Bronx, N.Y., assignor to Memorial Hospital for Cancer and Allied Diseases, New York, N.Y.
Continuation of Ser. No. 454,774, Dec. 30, 1982, abandoned. This application Oct. 4, 1985, Ser. No. 785,095
Int. Cl.⁴ H03H 19/00
U.S. Cl. 364—825   13 Claims

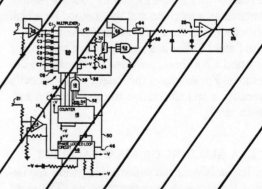

1. A sampled N-path filter, comprising:
(a) an N-path filter having N filter sections and means for connecting each filter section periodically into an input signal path through the N-path filter; and
(b) sampling means connected to the input signal path through the N-path filter during only a portion of each period each filter section of the N-path filter is connected into the input signal path through the N-path filter for

Fig. 4-1. Official Gazette entry: The Bracewell Transform.

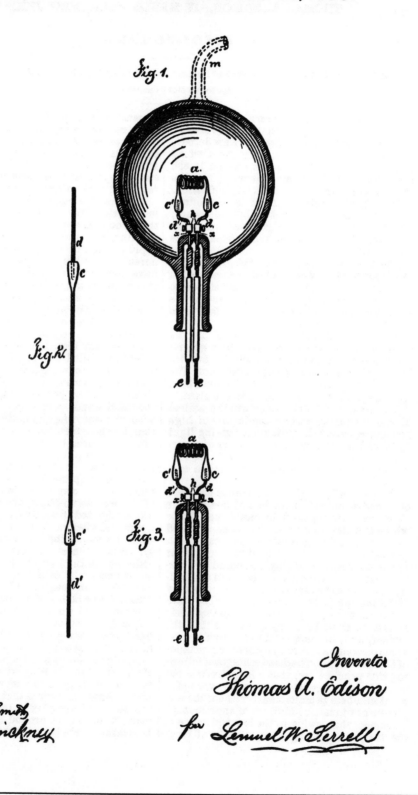

T. A. EDISON.
Electric-Lamp.

No. 223,898.   Patented Jan. 27, 1880.

Fig. 1.

Fig. 2.

Fig. 3.

Witnesses
Chas H Smith
Geo. T. Pinckney

Inventor
Thomas A. Edison
for Lemuel W. Serrell

Fig. 4-2. Patent: Edison's Electric Lamp (light bulb).

UNITED STATES PATENT OFFICE.

THOMAS A. EDISON, OF MENLO PARK, NEW JERSEY

ELECTRIC LAMP.

SPECIFICATION forming part of Letters Patent No. 223,898, dated January 27, 1880.

Application filed November 4, 1879.

To all whom it may concern:

Be it known that I, THOMAS ALVA EDISON, of Menlo Park, in the State of New Jersey, United States of America, have invented an 5 Improvement in Electric Lamps, and in the method of manufacturing the same, (Case No. 186,) of which the following is a specification.

The object of this invention is to produce electric lamps giving light by incandescence, 10 which lamps shall have high resistance, so as to allow of the practical subdivision of the electric light.

The invention consists in a light-giving body of carbon wire or sheets coiled or arranged in 15 such a manner as to offer great resistance to the passage of the electric current, and at the same time present but a slight surface from which radiation can take place.

The invention further consists in placing 20 such burner of great resistance in a nearly-perfect vacuum, to prevent oxidation and injury to the conductor by the atmosphere. The current is conducted into the vacuum-bulb through platina wires sealed into the glass.

25 The invention further consists in the method of manufacturing carbon conductors of high resistance, so as to be suitable for giving light by incandescence, and in the manner of securing perfect contact between the metallic con-30 ductors or leading-wires and the carbon conductor.

Heretofore light by incandescence has been obtained from rods of carbon of one to four ohms resistance, placed in closed vessels, in 35 which the atmospheric air has been replaced by gases that do not combine chemically with the carbon. The vessel holding the burner has been composed of glass cemented to a metallic base. The connection between the lead-40 ing wires and the carbon has been obtained by clamping the carbon to the metal. The leading-wires have always been large, so that their resistance shall be many times less than the burner, and, in general, the attempts of pre-45 vious persons have been to reduce the resistance of the carbon rod. The disadvantages of following this practice are, that a lamp having but one to four ohms resistance cannot be worked in great numbers in multiple arc without the em-50 ployment of main conductors of enormous dimensions; that, owing to the low resistance of the lamp, the leading-wires must be of large

dimensions and good conductors, and a glass globe cannot be kept tight at the place where the wires pass in and are cemented; hence the 55 carbon is consumed, because there must be almost a perfect vacuum to render the carbon stable, especially when such carbon is small in mass and high in electrical resistance.

The use of a gas in the receiver at the at-60 mospheric pressure, although not attacking the carbon, serves to destroy it in time by "air-washing," or the attrition produced by the rapid passage of the air over the slightly-coherent highly-heated surface of the carbon. I 65 have reversed this practice. I have discovered that even a cotton thread properly carbonized and placed in a sealed glass bulb exhausted to one-millionth of an atmosphere offers from one hundred to five hundred ohms resistance to the 70 passage of the current, and that it is absolutely stable at very high temperatures; that if the thread be coiled as a spiral and carbonized, or if any fibrous vegetable substance which will leave a carbon residue after heating in a 75 closed chamber be so coiled, as much as two thousand ohms resistance may be obtained without presenting a radiating-surface greater than three-sixteenths of an inch; that if such fibrous material be rubbed with a plastic com-80 posed of lamp-black and tar, its resistance may be made high or low, according to the amount of lamp-black placed upon it; that carbon filaments may be made by a combination of tar and lamp-black, the latter being pre-85 viously ignited in a closed crucible for several hours and afterward moistened and kneaded until it assumes the consistency of thick putty. Small pieces of this material may be rolled out in the form of wire as small as seven 90 one-thousandths of a inch in diameter and over a foot in length, and the same may be coated with a non-conducting non-carbonizing substance and wound on a bobbin, or as a spiral, and the tar carbonized in a closed cham-95 ber by subjecting it to high heat, the spiral after carbonization retaining its form.

All these forms are fragile and cannot be clamped to the leading-wires with sufficient force to insure good contact and prevent heat-100 ing. I have discovered that if platinum wires are used and the plastic lamp-black and tar material be molded around it in the act of carbonization there is an intimate union by com-

material be molded around it in the act of carbonization there is an intimate union by combination and by pressure between the carbon and platina, and nearly perfect contact is obtained without the necessity of clamps; hence the burner and the leading-wires are connected to the carbon ready to be placed in the vacuum-bulb.

When fibrous material is used the plastic lamp-black and tar are used to secure it to the platina before carbonizing.

By using the carbon wire of such high resistance I am enabled to use fine platinum wires for leading-wires, as they will have a small resistance compared to the burner, and hence will not heat and crack the sealed vacuum-bulb. Platina can only be used, as its expansion is nearly the same as that of glass.

By using a considerable length of carbon wire and coiling it the exterior, which is only a small portion of its entire surface, will form the principal radiating-surface; hence I am able to raise the specific heat of the whole of the carbon, and thus prevent the rapid reception and disappearance of the light, which on a plain wire is prejudicial, as it shows the least unsteadiness of the current by the flickering of the light; but if the current is steady the defect does not show.

I have carbonized and used cotton and linen thread, wood splints, papers coiled in various ways, also lamp-black, plumbago, and carbon in various forms, mixed with tar and kneaded so that the same may be rolled out into wires of various lengths and diameters. Each wire, however, is to be uniform in size throughout.

If the carbon thread is liable to be distorted during carbonization it is to be coiled between a helix of copper wire. The ends of the carbon or filament are secured to the platina leading-wires by plastic carbonizable material, and the whole placed in the carbonizing-chamber. The copper, which has served to prevent distortion of the carbon thread, is afterward eaten away by nitric acid, and the spiral soaked in water, and then dried and placed on the glass holder, and a glass bulb blown over the whole, with a leading-tube for exhaustion by a mercury-pump. This tube, when a high vacuum has been reached, is hermetically sealed.

With substances which are not greatly distorted in carbonizing, they may be coated with a non-conducting non-carbonizable substance, which allows one coil or turn of the carbon to rest upon and be supported by the other.

In the drawings, Figure 1 shows the lamp sectionally. a is the carbon spiral or thread. $c\ c'$ are the thickened ends of the spiral, formed of the plastic compound of lamp-black and tar. $d\ d'$ are the platina wires. $h\ h$ are the clamps, which serve to connect the platina wires, cemented in the carbon, with the leading-wires $x\ x$, sealed in the glass vacuum-bulb. $e\ e$ are copper wires, connected just outside the bulb to the wires $x\ x$. m is the tube (shown by dotted lines) leading to the vacuum-pump, which, after exhaustion, is hermetically sealed and the surplus removed.

Fig. 2 represents the plastic material before being wound into a spiral.

Fig. 3 shows the spiral after carbonization, ready to have a bulb blown over it.

I claim as my invention—

1. An electric lamp for giving light by incandescence, consisting of a filament of carbon of high resistance, made as described, and secured to metallic wires, as set forth.

2. The combination of carbon filaments with a receiver made entirely of glass and conductors passing through the glass, and from which receiver the air is exhausted, for the purposes set forth.

3. A carbon filament or strip coiled and connected to electric conductors so that only a portion of the surface of such carbon conductors shall be exposed for radiating light, as set forth.

4. The method herein described of securing the platina contact-wires to the carbon filament and carbonizing of the whole in a closed chamber, substantially as set forth.

Signed by me this 1st day of November, A. D. 1879.

THOMAS A. EDISON.

Witnesses:
S. L. GRIFFIN,
JOHN F. RANDOLPH.

It is found that the following certificate has been attached to Letters Patent granted to Thomas A. Edison for improvement in "Electric Lamps," No. 223,898, dated January 27, 1880:

DEPARTMENT OF THE INTERIOR,

UNITED STATES PATENT OFFICE,

WASHINGTON, D. C., *December 18, 1883.*

In compliance with the request of the party in interest Letters Patent No. 223,898, granted January 27, 1880, to Thomas A. Edison, of Menlo Park, New Jersey, for an improvement in "Electric Lamps," is hereby limited so as to expire at the same time with the patent of the following-named, having the shortest time to run, viz.: British patent, dated November 10, 1879, No. 4,576; Canadian patent, dated November 17, 1879, No. 10,654; Belgian patent, dated November 29, 1879, No. 49,884; Italian patent, dated December 6, 1879, and French patent, dated January 20, 1880, No. 133,756.

It is hereby certified that the proper entries and corrections have been made in the files and records of the Patent Office.

This amendment is made that the United States patent may conform to the provisions of section 4887 of the Revised Statutes.

[SEAL.]

BENJ. BUTTERWORTH,
Commissioner of Patents.

Approved:

M. L. JOSLYN,
Acting Secretary of the Interior.

Now, in compliance with the request of the parties in interest, said certificate is hereby *canceled* and proper entries and corrections have been made in the files and records of the Patent Office.

In testimony whereof I have hereunto set my hand and caused the seal of the Patent Office to be affixed, this 15th day of March, 1893.

W. E. SIMONDS,
Commissioner of Patents.

Approved:

CYRUS BUSSEY,
Assistant Secretary of the Interior.

DEPARTMENT OF THE INTERIOR,

UNITED STATES PATENT OFFICE,

WASHINGTON, D. C., *December 18, 1883.*

In compliance with the request of the party in interest, Letters Patent No. 223,898, granted January 27, 1880, to Thomas A. Edison, of Menlo Park, New Jersey, for an improvement in "Electric Lamps," is hereby limited so as to expire at the same time with the patent of the following named, having the shortest time to run, vis: British Patent dated November 10, 1879, No. 4,576; Canadian Patent dated November 17, 1879, No. 10,654; Belgian Patent dated November 29, 1879, No. 49,884; Italian Patent dated December 6, 1879; and French Patent dated January 20, 1880, No. 133,756;

It is hereby certified that the proper entries and corrections have been made in the files and records of the Patent Office.

This amendment is made that the United States Patent may conform to the provisions of Section 4887 of the Revised Statutes.

BENJ. BUTTERWORTH,
Commissioner of Patents.

Approved:

M. L. JOSLYN,
Acting Secretary of the Interior.

United States Patent

[11] **3,630,430**

[72] Inventor **Glenn E. Struble**
Fairfield, Ohio

[21] Appl. No. **61,621**
[22] Filed **Aug. 6, 1970**
[45] Patented **Dec. 28, 1971**
[73] Assignee **Diamond International Corporation**
New York, N.Y.

[54] **QUICKLY ERECTED SCOOP-TYPE CARTON**
6 Claims, 4 Drawing Figs.

[52] U.S. Cl... **229/16 B,**
229/1.5 B, 229/41 B, 229/41 D
[51] Int. Cl.. **B65d 5/36**
[50] Field of Search.. 229/21, 1.5
B, 16 R, 41 R, 41 B, 41 C, 41 D; 294/55

[56] **References Cited**
UNITED STATES PATENTS

904,050	11/1908	Crawford	229/21
1,034,522	8/1912	Shaw	229/1.5 B UX
2,078,038	4/1937	Stephens	229/16 R X
2,337,199	12/1943	Holy	229/16 R
2,385,898	10/1945	Waters	229/1.5 B X

Primary Examiner—Donald F. Norton
Attorney—Karl W. Flocks

ABSTRACT: A box for quick erection in the shape of a scoop made with application of glue in parallel strips parallel to the blank edges and with an arcuate bottom having curved score lines and tapered sides to the box in assembled form.

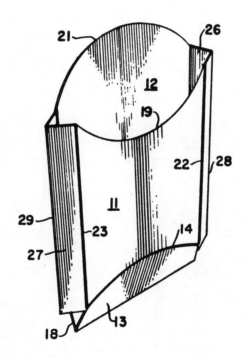

Fig. 4-3. Patent: Quickly Erected Scoop-Type Carton (McDonald's extra large fries box).

FIG.1.

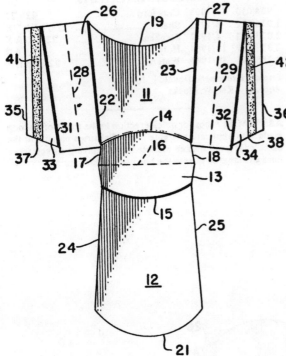

FIG.2.

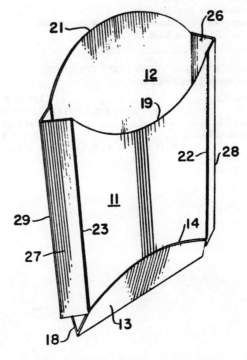

FIG.3.

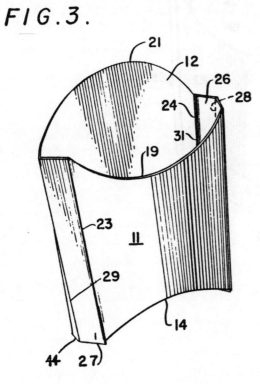

FIG.4.

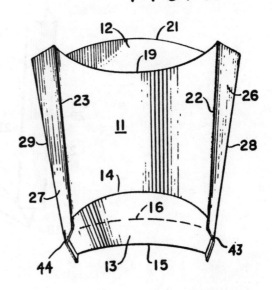

INVENTOR

GLENN E. STRUBLE

BY KARL W. FLOCKS

ATTORNEY

1

QUICKLY ERECTED SCOOP-TYPE CARTON

BACKGROUND OF THE INVENTION

The present invention relates to cartons, and more particularly to folding cartons of the type including scored fold lines on the end and bottom wall panels so that they may be expanded readily from flat to setup position.

In restaurants, especially of the carryout type there is a need for open end cartons that can be simply and inexpensively manufactured and then can be stored in flattened condition until ready for use. These cartons must then be able to be assembled with the very minimum of effort and time and then serve a dual purpose of scooping up the items and then holding these items until they are consumed.

Such a food item for which the carton of the present invention is most useful is in the serving of french fried potatoes. It has been found most desirable if such a carton could be picked up and be assembled merely by the use of one hand and immediately be used to scoop up the food item and be handed to the customer without further effort or utensils being used.

Since the food item would of course be spilled and wasted if the box should collapse while being used by the customer it was important that although the carton be easy to assemble, the assembled carton must positively hold its shape and not tend to reassume its flattened position.

Therefore, the present invention is a carton which has the combined attributes of inexpensive manufacture, quick and easy opening into a positive locked assembled carton, and shaped for scooping.

SUMMARY OF THE INVENTION

The present invention is directed to a carton blank having parallel edges along which parallel lines of glue are applied, and arcuate bottom formed by curved score lines, glue flaps each with a sharply beveled edge to accommodate the curved score lines, and the end panels shaped for scoop and receiving tray construction.

The present invention is also directed to the carton formed from said blank forming a scoop flap with one end panel and inwardly bowed bottom having edges contoured to the side panels by score lines and free edges contoured to the end wall panels.

BRIEF DESCRIPTION OF THE DRAWINGS

FIG. 1 illustrates the blank of the carton in accordance with the present invention;

FIG. 2 shows the blank of FIG. 1 assembled and in a partially expanded condition;

FIG. 3 is a perspective view of the carton of FIG. 2 in setup condition ready for the scooping and insertion of a product therein; and

FIG. 4 is another perspective view of the carton of FIG. 3 in setup condition, as seen from a side-bottom view.

DESCRIPTION OF THE PREFERRED EMBODIMENT

Referring first to FIG. 1 the carton blank there illustrated includes a pair of sidewall panels 11 and 12 each having a shorter edge attached to opposite sides of a bottom panel 13 along arcuate score lines 14 and 15 respectively. Bottom panel 13 is bisected by score line 16, and the edges 17, 18 of bottom panel 13 which are not connected to side panels 11 and 12 are formed as inwardly sloping from the bisecting score line 16 toward, and connected to the ends of, arcuate score lines 14 and 15.

Arcuate score lines 14 and 15 are curved so as to produce concave edges on side panels 11 and 12 respectively. The opposite edge of sidewall panel 11 is also inwardly curved to form a concave edge 19 to that panel. The form of sidewall panel 12 differs in that its opposite edge is curved outwardly to form a convex edge 21. Sidewall panel 11 has score lines 22 and 23 extending laterally and outwardly tapering between arcuate score line 14 and concavely curved edge 19. Sidewall

2

panel 12 also has outwardly tapering edges 24 and 25 extending from arcuate score line 15 to the ends of convex edge 21. End wall panels 26 and 27 extend laterally along score lines 22 and 23 respectively and are thereby connected to sidewall panel 11. End wall panels 26 and 27 are each bisected by a score line 28 and 29 respectively. Along their opposite lateral edges at score lines 31 and 32 respectively are attached glue flaps 33 and 34 respectively. Glue flaps 33 and 34 have outer free edges 35 and 36 respectively which are parallel to each other and which have their bottom edges 37 and 38 respectively forming a beveled edge between the outer free edges 35 and 36 and score lines 31 and 32 respectively. This beveled shape is of a sufficient angle to exceed the angle of the arcuate edge 15 so as not to extend below that arcuate edge when the carton is in a glued, and either folded or extended position.

Glue lines 41 and 42 are substantially parallel to each other and to outer free edges 35 and 36 respectively of glue flaps 33 and 34 since parallel application of glue on such a blank is much easier, requiring simpler gluing machinery using the technique and equipment customarily used in the folding carton industry.

End wall panels 26 and 27 extend below the intersections of score lines 22 and 23 respectively with score line 14, so as to form legs 43 and 44 respectively which are used to support the erected carton in an upright position.

In order to better illustrate the folded flap position of a carton of the present invention, FIG. 2 shows the carton in a partially folded position. This carton which is easily formed by conventional folding machinery has been formed by first folding the carton blank along score line 16 and after applying glue lines 41 and 42, then folding the blank along score lines 28 and 29, thereby forming the carton in its closed and flattened position. In this position score lines 22 and 23 lies substantially parallel and immediately adjacent to score lines 31 and 32, respectively.

When the carton is to be used it will be shaped to its useful form as shown in FIGS. 3 and 4. This is easily accomplished by pressing with the fingers along score lines 28, 29 and 16. Due to the arcuate score lines 14 and 15 along bottom panel 13 the carton will be formed in a positive manner so as not to return to a flattened position and thereby spill the contents therein.

It will be noted as clearly illustrated in FIG. 4 that the end panels are tapered outwardly due to the tapered form of the respective edges 22, 23, 24 and 25 of end wall panels 11 and 12. This outward taper in cooperation with the convex edge 21 of sidewall panel 12 and concave edge 19 of sidewall panel 11 form a scoop allowing better use of the carton of the present invention. By having this scoop formed in the manner shown, the carton in its use at food carryout shops allows the scooping up of a serving of french fries or the like without the use of additional utensils to fill the carton, speeding the operation since only one hand is needed for both the forming of the carton and the scooping of the food product into it. Since the picking up of a serving is so easily accomplished such foods may remain in their respective pot or pan and thus be kept hot until immediately before serving to the customer. The positive action keeping the carton open through its use of arcuate score lines allows use of the carton in carryout food shops with less chance of spilling of the contents and easier handling of the carton while eating and consuming its contents. In addition, the convex versus concave shapes in cooperation with each other and the tapered end panels eases the scooping action by providing a wide mouth end for picking up and serving of the food product.

It will be apparent that modifications in accordance with the present invention can be made by those skilled in the art without departing from the spirit thereof and it is equally apparent that the assembly involving the application of glue and folding of the carton blank may be rearranged in the order of accomplishing these steps without departing from the scope of the invention.

It will be obvious to those skilled in the art that various changes may be made without departing from the scope of the invention and the invention is not to be considered limited to what is shown in the drawings and described in the specification.

What is claimed is:

1. A carton formed from a single blank of foldable paperboard comprising

a bottom panel having arcuate scored edges and bisected by a score line,

a pair of sidewall panels, each hingedly connected by one of said arcuate scored edges to said bottom panel,

a pair of end wall panels hingedly attached to opposite lateral edges of one of said sidewall panels by score lines tapering outwardly from said bottom panel connection,

and a glue flap hingedly attached to each of said end wall panels opposite said tapering score line by a score line substantially parallel to said tapering score line when the carton is in folded flattened condition but tapering outwardly from said tapering score line and upwardly from the bottom of said end wall panel in other than flattened condition

whereby said glue flaps overlap the other of said pair of sidewall panels when in a folded and glued position.

2. The carton of claim 1, further characterized by said bottom panel having opposite free edges inclined inward from their intersection with said score line bisecting said bottom panel to their intersection with said arcuate scored edges.

3. The carton of claim 1, further characterized by said pair of sidewall panels each having an outer free edge opposite said arcuate scored edge,

one of said free edges having an arcuate concave shape and the other of said free edges having an arcuate convex shape.

4. The carton of claim 1 further characterized by each of said end wall panels bisected by a score line from top to bottom.

5. The carton of claim 1, further characterized by said glue flaps each having a beveled edge extending from said score line attaching each said flap to its adjacent end wall panel, to the outer free edge of said glue flap,

said beveled edge having a bevel thereto extending said edge above the arcuate score line of said overlapped sidewall panel.

6. The carton of claim 1, further characterized by each of said end wall panels extending below said bottom panel when the carton is in erected position, forming legs below said bottom panel.

* * * * *

WHAT IS AN IMPROVEMENT OF A PREVIOUSLY EXISTING PROCESS, MACHINE, MANUFACTURE OR COMPOSITION OF MATTER?

Any physical difference in the construction of a machine or manufacture or change in the steps of a method that results in improved properties can be an improvement invention.

Improvement inventions show how to take something old and make it better. If the something that is old is from an expired patent or is otherwise not subject to someone else's patent rights, then the owner of the improvement patent gains the exclusive right to use the improved invention. If the improvement pertains to an invention in a non-expired patent, then the owner of the improvement patent will not be able to use his rights without permission of the previous owner. However, the previous owner won't be able to use the improvement without permission of the improvement patent owner. The basis for a mutually beneficial swapping of rights thus exists.

Requirement B: Utility

The second requirement for a patent application is that it must disclose an invention that has utility, that is, is useful for something. If the detailed description section of your patent application contains the answer to the question "What is your invention good for?", then the utility requirement is probably satisfied.

There are several other ways an invention might fail to have utility in the eyes of the Patent Office. These include the grounds of (1) inoperativeness; (2) involving perpetual motion; (3) frivolous; (4) fraudulent; (5) and against public policy.

INOPERATIVENESS

Inoperative inventions are those that do not do what they are supposed to do. If, based on what the patent examiner can see from your application, the invention will not work, the utility requirement is not met. This frequently happens because the applicant has presented an improper description or drawing of the invention. If a critical part is omitted from the description, the resulting application winds up describing something that will not work.

Inoperativeness might also be the case if the application fails to state what the invention is good for. If the application describes a new chemical compound but fails to state that the compound is good as an antibiotic or detergent, then the application won't fly. The utility requirement also will not be satisfied if the application states that the sole use of the new compound is to make some other compound, the usefulness of which is also unknown.

PERPETUAL MOTION

Inventions involving perpetual motion are considered inoperative because the Patent Office feels that these violate the laws of thermodynamics. If your invention involves what appears to be perpetual motion, you might be required to submit a prototype before a patent will be granted. You can avoid problems with this type of invention by including in your patent application a discussion of the theory of operation. A nuclear reactor, to an outside observer, appears to be a perpetual motion machine because it can put out energy for long periods of time without fuel input. However, this illusion disappears once the theory of operation is explained.

FRIVOLOUS

Frivolous inventions are those that do not accomplish technologically useful results. If the invention is a pretty, new shape for a lamp base or a pyramidal device for improving the local karma, it could be rejected as frivolous. For the lamp base, at least, you might consider putting a copyright on it as a work of sculpture or else apply for a different type of patent: a design patent. Design patents, which protect ornamental designs as opposed to useful inventions, are not covered in this book.

FRAUDULENT

Inventions that are fraudulent and against public policy are those that have utility only for "evil" purposes. Devices for jamming police radar, for receiving cable TV without paying for it, or something promoting gambling will not qualify as useful. On the other hand, devices for jamming the radar of attacking Russian bombers and descrambling cable TV for authorized customers are not fraudulent or against public policy. Write your application carefully! Because most state governments now operate lotteries, you might be able to persuade a patent examiner that some gambling devices also have utility. In any case, be sure to mention at least one morally correct use for your invention.

Requirement C: Newness

There are two ways that an invention can fail to be new. Either someone invented it before, or you invented it first but waited too long before getting your patent application filed. Prior invention by another is called *anticipation*. Waiting too long to file results in a *statutory bar* because the patent statutes bar the grant of patents to those who wait too long before filing.

WHEN IS AN INVENTION ANTICIPATED?

Anticipation. An invention is anticipated if someone else already has a U.S. or foreign patent granted at the time you conceive the idea. You're also out of luck if, at the time you conceive the invention, someone else has already filed for a patent in the U.S. or described it in print anywhere in the world. Alternatively, you can be out of luck if the invention is already known or used anywhere in the United States.

After you file a patent application, the patent examiner will explore the prior art, that is, all of the records and files of the Patent Office technical library, in order to see if the invention is anticipated. Because it costs money to file for a patent and it takes a lot of work to write the application, wouldn't it be nice to find out whether you are anticipated before wasting all this effort? For this, you need to do a patent search. A patent search can also help you write a patent application that is more likely to succeed because you will obtain example patents from the same area of technology as your invention. The writing style and approach of these patents can be usefully adapted to your own application. (Chapter 3 describes how to do a patent search.)

WHEN DOES A STATUTORY BAR APPLY?

Statutory Bar. Even if your invention is not anticipated, it can still fail to be new due to some act on your part—a statutory bar. A statutory bar arises if

(1) The invention was described in a printed publication written by you or anyone else more than one year before the filing date of the patent application; or

(2) You file a patent application in a foreign country more than 12 months before filing in the U.S., and the foreign country issues a patent before your U.S. filing date; or

(3) You or someone else put the invention into public use or on sale more than 12 months before the filing date.

Statutory bars are intended to encourage inventors to file for patents quickly. You can avoid statutory bar problems, at least those arising from your own activity, by filing your patent application promptly after coming up with a good idea. You should be especially wary of the public use or sale type of bar. The use need not be very public. For example, in one famous case, the inventor of a new corset allowed a female acquaintance to use it for too long a time before filing his application. Because the woman had appeared in public, this was found to be a public use of the corset. In another case, it was held that marketing products made with a secret process resulted in a bar to a patent on the process—even though the public could not figure out the process by merely looking at the product.

A copy of the U.S. Supreme Court case concerning the corset, *Egbert vs. Lippman,* appears in Fig. 4-4. Notice that the Court refers to a two-year time limit for public use, after which a patent is barred. Under current law, the time limit is one year.

Statutory bar problems are likely to arise during alpha or beta testing of computer or other devices by potential users. Although purely experimental use of an invention does not result in a bar, the standards for experimental use are very strict. There cannot be the slightest taint of commercialism in it . . . as there often is when a potential customer is trying out a new product.

The best cure for statutory bar problems is prevention. If you are reasonably sure that your concept will work, do not hesitate to file for a patent even if you haven't built and tested the thing yet. Bell filed for his telephone patent before he made his famous first call to Watson. The Wright air-

plane patent was filed on March 23, 1903, but the first flight at Kitty Hawk did not happen until December. This type of filing is called a *constructive reduction to practice.* It gives the inventor a definite priority date for his invention and prevents any future problems with statutory bars. If the invention works when it is actually built, the patent is valid. In Bell's case, the instructions in his early-filed patent, No. 174,456, were sufficient to teach an ordinary skilled mechanic how to make and use a working telephone, even though Bell had not yet built one. Likewise, the airplane, when built as described in the Wright's application, did fly.

Requirement D: Originality

The patent examiner, in his search of the prior art, often fails to find a prior invention that is exactly the same. Therefore, he cannot refuse to grant a patent upon the basis of anticipation. However, the law also prohibits issuance of a patent if the invention is considered to be "obvious" in view of what is to be found in previous technical literature and patents, that is, the prior art. Whether or not an invention is obvious depends on the answer to this question: "Would a person who is reasonably knowledgeable in this type of technology, upon a reading of what is taught in the prior art, be led to the same inventive concept?" If the answer to this question is yes, then obviousness exists, and the patent application can be rejected.

Anticipation exists when the patent examiner can come up with an old patent or article that describes your invention exactly. This means that the complete concept is shown within the four corners of a single document. However, to show obviousness, the examiner is permitted to combine the teachings of several prior art documents in order to prove that your concept is not really new. The standard for determining obviousness can be re-stated this way: "Would the reasonably knowledgeable person in this technology, upon reading these documents in conjunction with one another, be led to your inventive concept?"

To satisfy this standard, the examiner must show (1) that there is some motivation or reason to combine the teachings of the documents in the exact way required to arrive at your invention; and (2) that it is not reasonable to combine the various documents in other ways that result in the teaching of some other concept. Examples of typical rejections for anticipation and obviousness (and ways to respond to them) are given in Chapter 7.

Requirement E: Completeness of Description

The patent application must clearly describe how to make. and use the invention. Drawings must be included if possible. This requirement means that you must tell all regarding your invention. You must describe the best version of your invention of which you have knowledge—this means the version you consider to be of most economic value. If it is pos-

thirds of the land in dispute. If that be not in legal effect, a finding that defendants were in possession of the entire land, there is a finding that defendants respectively, claimed title to the several tracts in controversy. The verdict describes by metes and bounds each tract embraced in the suit, giving the name of each defendant by whom it is claimed and finding for defendants as to two thirds, undivided, of the respective tracts. It then proceeds to find as "to the remaining one third of the lands hereinbefore excepted, and *claimed by said defendants.*" Although the verdict does not state, in terms, that the defendants were in possession, it does state that they claimed the lands in dispute. And that seems to be sufficient under the local law. An reference to the case of *Southgate* v. *Walker,* 2 W. Va., 427, it is sufficient to say that it related to an action of ejectment commenced in 1848, before the adoption of the above-recited provision. We are referred to no decision of the state court in conflict with the construction we have given to that provision.

We deem it unnecessary to comment upon any other objections urged against the special verdict. *There is no error in the judgment, and it is affirmed.*

True copy. Test:
James H. McKenney, Clerk, Sup. Court, U.S.

Cited—107 U. S., 500.

FRANCES LEE EGBERT, Exrx. of SAMUEL H. BARNES, Deceased, *Appt.*,

v.

PHILLIPP LIPPMANN ET AL.

(See S. C., 14 Otto, 333–339.)

Public use of patented invention—what is—void letters patent.

1. To constitute the public use of an invention, it is not necessary that more than one of the patented articles should be publicly used; such use may be of an invention, by its character, only capable of being used where it cannot be seen nor observed by the public eye.
2. Whether the use of an invention is public or private does not necessarily depend upon the number of persons to whom its use is known.
3. Re-issued letters patent to Frances Lee Barnes, executrix, for an improvement in corset springs, are void on account of two years' public use, by the consent and allowance of the inventor, before he made application for letters patent.

[No. 89.]

Argued Nov. 11, 14, 1881. Decided Dec. 12, 1881.

APPEAL from the Circuit Court of the United States for the Southern District of New York.

The case is stated by the court.

Messrs. **J. C. Clayton,** *Geo. Gifford* and **Anthony L. Keasbey,** for appellant.

Mr. **John B. Staples,** for appellees.

Mr. Justice **Woods** delivered the opinion of the court:

This suit was brought for an alleged infringement of the complainant's re-issued letters patent, No. 5216, dated January 7, 1873, for an infringement in corset springs.

The original patent bore date July 17, 1866, and was issued to Samuel H. Barnes.

The re-issue was made to the appellant,

See 14 OTTO.

Frances Lee Egbert, executrix of the original patentee.

The specifications for the re-issue declared:

" This invention consists in forming the springs of corsets of two or more metallic plates, placed one upon another, and so connected as to prevent them from sliding off each other laterally or edgewise, and at the same time admit of their playing or sliding upon each other in the direction of their length or longitudinally, whereby their flexibility and elasticity are greatly increased, while at the same time much strength is obtained."

The second claim was as follows:

"A pair of corset springs, each member of the pair being composed of two or more metallic plates, placed one on another and fastened together at their centers, and so connected at or near each end that they can move or play on each other in the direction of their length."

The bill of complaint alleged that Barnes was the original and first inventor of the improvement covered by the re-issued letters patent, and that it had not, at the time of his application for the original patent, been for more than two years in public use or on sale, with his consent or allowance.

The answer took issue on this averment and also denied infringement. The circuit court dismissed the bill and the complainant appealed to this court.

As to the second defense above mentioned, it is sufficient to say that the evidence establishes beyond controversy the infringement by defendants of the second claim of the re-issued letters patent above set forth.

We have, therefore, to consider whether the defense of the public use of the patented invention, with the consent of the inventor, for more than two years prior to his application for the original patent, is sustained by the testimony in the record.

The 6th, 7th and 15th sections of the Act of July 4, 1836, 5 Stat. at L., 117, as qualified by the 7th section of the Act of March 3, 1839, 5 Stat. at L., 353, were in force in 1866, when Barnes applied for his patent.

The effect of these sections is to render letters patent invalid, if the invention which they cover was in public use, with the consent and allowance of the inventor, for more than two years prior to his application for a patent. Since the passage of the Act of 1839 it has been strenuously contended that a public use of an invention, for more than two years before the application for a patent, even without the consent and allowance of the inventor, rendered the patent therefor void.

It is unnecessary in this case to decide this question, for the alleged use of the invention covered by the patent to Barnes is conceded to have been with his express consent.

The evidence on which the defendants rely to establish a prior public use of the invention, consists mainly of the testimony of the complainant herself, who is the executrix of the original patentee.

She testifies that Barnes invented the improvement covered by his patent between January and May, 1855; that between the dates named the witness and her friend, Miss Cugier, were complaining of the breaking of their corset steels. Barnes, who was present, and was an

Fig. 4-4. U.S. Supreme Court Opinion, October Term, 1981: Egbert V. Lippman.

intimate friend of the witness, said he thought he could make her a pair that would not break. At their next interview he presented her with a pair of corset steels which he himself had made. The witness wore these steels a long time. In 1858 Barnes made and presented to her another pair, which she also wore a long time. When the corsets in which these steels were used wore out, the witness ripped them open and took out the steels and put them in new corsets. This was done several times.

It is admitted and, in fact, is asserted, by complainant that these steels embodied the invention afterward patented by Barnes and covered by the re-issued patent on which this suit is brought.

Joseph H. Sturgis, another witness for complainant, testifies that in 1863 Barnes spoke to him about two inventions made by himself, one of which was a corset steel, and that he went to the house of Barnes to see them. Before this time, and after the transactions testified to by the complainant, Barnes and she had intermarried. Barnes said his wife had a pair of steels made according to his invention in the corsets which she was then wearing, and if she would take them off he would show them to witness. Mrs. Barnes went out and returned with a pair of corsets and a pair of scissors and ripped the corsets open and took out the steels. Barnes then explained to witness how they were made and used.

This is the evidence presented by the record, on which defendants rely to establish the public use of the invention by the patentee's consent and allowance.

The question for our decision is, whether this testimony shows a public use within the meaning of the statute.

We observe in the first place that to constitute the public use of a patent it is not necessary that more than one of the patented articles should be publicly used. The use of a great number may tend to strengthen the proof of public use, but one well defined case of public use is just as effectual to annul the patent as many.

For instance, if the inventor of a mower, a printing press or a railway car, makes and sells only one of the articles invented by him, and allows the vendee to use it for two years, without restriction or limitation, the use is just as public as if he had sold and allowed the use of a great number.

We remark, secondly, that, whether the use of an invention is public or private, does not necessarily depend upon the number of persons to whom its use is known. If an inventor, having made his device, gives or sells it to another, to be used by the donee or vendee, without limitation or restriction, or injunction of secrecy, and it is so used, such use is public, within the meaning of the statute, even though the use and knowledge of the use may be confined to one person.

We say thirdly, that some inventions are by their very character only capable of being used where they cannot be seen or observed by the public eye. An invention may consist of a lever or spring, hidden in the running gear of a watch, or of a ratchet, shaft or cog-wheel covered from view in the recesses of a machine for spinning or weaving. Nevertheless, if its

inventor sells a machine of which his invention forms a part, and allows it to be used without restriction of any kind, the use is a public one, within the meaning of the law. So, on the other hand, a use necessarily open to public view, if made in good faith solely to test the qualities of the invention, and for the purpose of experiment, is not a public use within the meaning of the patent law. *Elizabeth* v. *Pavement Co.*, 97 U. S., 126 [XXIV., 1000]; *Shaw* v. *Cooper,* 7 Pet., 292.

Tested by these principles, we think the evidence of the complainant herself shows that there was a public use of the invention, covered by the original patent to Barnes, for more than two years before the application for the patent, and by his consent and allowance. He made and gave to the complainant two pairs of corset steels, constructed according to his device, one in 1855 and one in 1858. They were presented to her for use. He imposed no obligation of secrecy, or any condition or restriction whatever. They were not presented for the purpose of experiment or to test their qualities. No such claim is set up in the testimony of complainant. The invention was at the time complete, and there is no evidence that it was afterwards changed or improved. The donee of the steels used them for years for the purpose and in the manner designed by the inventor. They were not capable of any other use. She might have exhibited them to any person she pleased, or might have made other steels of the same kind, and used or sold them without violation of any condition or restriction imposed on her by the inventor.

According to the testimony of complainant, the invention was completed and put in use in 1855. The inventor slept on his rights for eleven years. The patent was not applied for till March, 1866. In the meantime, the invention had found its way into general, and almost universal use. A great part of the record is taken up with the testimony of the manufacturers and venders of corset steels, showing that before Barnes applied for his patent the principle of his device was almost universally used in the manufacture of corset steels. It is fair to presume that having learned from this general use that there was some value in his invention, Barnes attempted to resume, by an application for a patent, what by his acts he had clearly dedicated to the public.

"An abandonment to the public may be evinced by the conduct of the inventor at any time, even within the two years named in the law. The effect of the law is that no such consequence will necessarily follow from the invention being in public use or on sale with the inventor's consent and allowance, at any time within two years before his application, but that if the invention is in public use or on sale prior to that time, it will be conclusive evidence of abandonment, and the patent will be void." *Elizabeth* v. *Pavement Co., supra.*

We are of opinion that the defense of two years' public use, by the consent and allowance of the inventor, before he made application for his patent, is satisfactorily established by the evidence. *The decree of the Circuit Court is, therefore, affirmed.*

True copy. Test:
    James H. McKenney, Clerk, Sup. Court, U. S.

United States Patent [19]

Rachofsky

[11] **Patent Number:** **4,541,726**

[45] **Date of Patent:** **Sep. 17, 1985**

[54] **TWENTY-FIVE (25) HOUR CLOCK**

[76] Inventor: Morton Rachofsky, 5511 Stonegate Rd., Dallas, Tex. 75209

[21] Appl. No.: 650,730

[22] Filed: **Sep. 17, 1984**

[51] Int. Cl.⁴ G04B 19/04
[52] U.S. Cl. 368/80; 368/220; 368/228
[58] Field of Search 368/220, 221, 223, 228, 368/232, 76, 80, 82–84

[56] **References Cited**

U.S. PATENT DOCUMENTS

4,175,378 11/1979 Shelton 368/20

Primary Examiner—Vit W. Miska
Attorney, Agent, or Firm—David H. Judson

[57] **ABSTRACT**

A clock for representing twenty-five (25) simulated hours in a twenty-four (24) hour real-time day is provided. The clock preferably includes a clock face having indicia printed thereon at twenty-five (25) equally-spaced intervals, each of the intervals representing a simulated hour in a twenty-five (25) hour simulated day. The clock includes a conventional hour and minute hand for cooperating with the clock face to provide a time indication. A clock drive mechanism simultaneously drives the minute hand around the clock face, and the hour hand between spaced intervals, in 1/25th of the twenty four hour real-time day. The clock may also be provided with a second hand driven by the clock drive mechanism around the clock face in 24/25th's of a real-time minute.

7 Claims, 6 Drawing Figures

Fig. 4-5. Patent: 25-Hour Clock.

FIG. 1

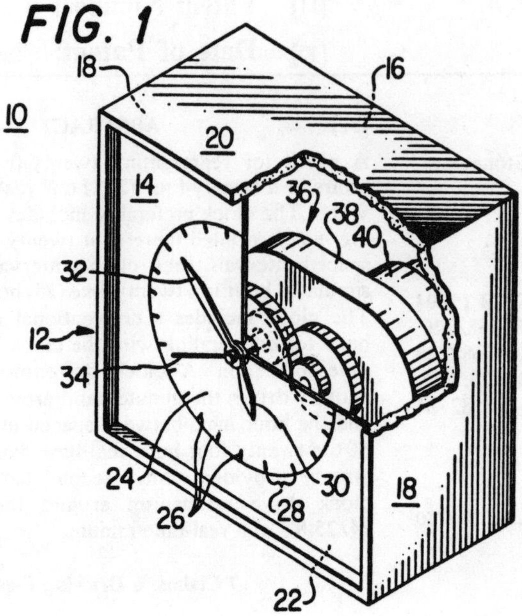

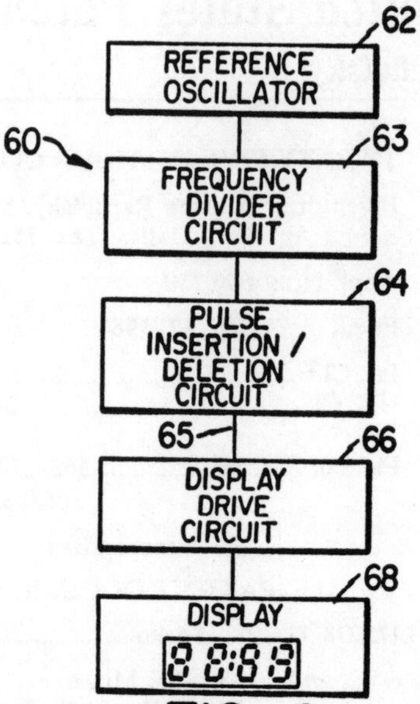

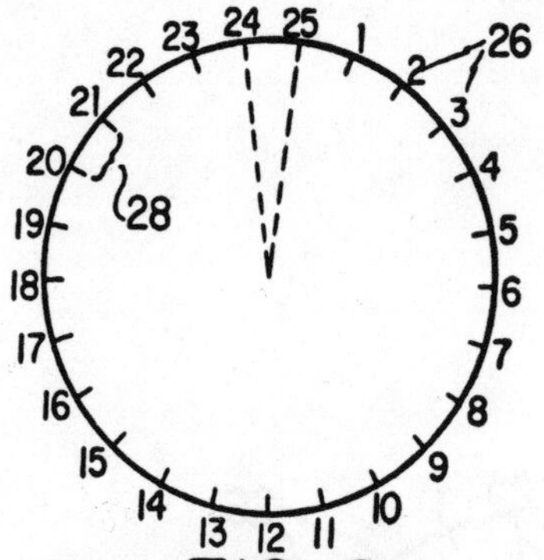

FIG. 2a

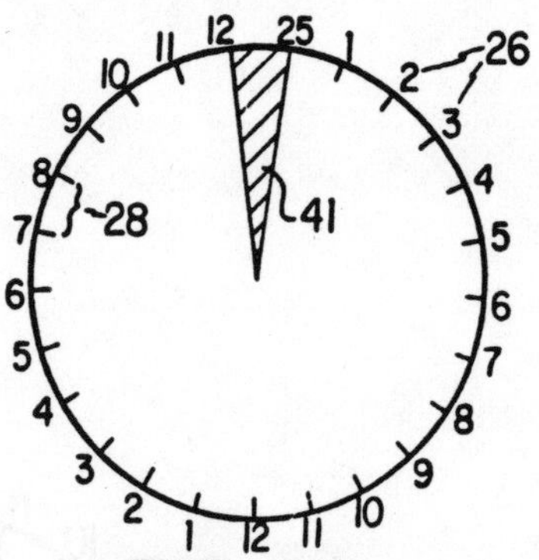

FIG. 2b

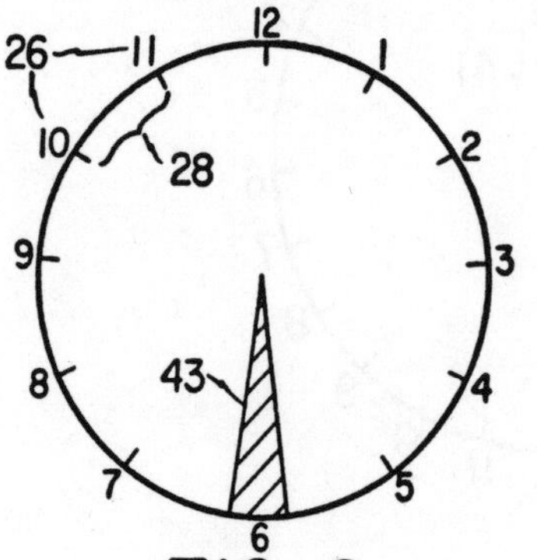

FIG. 2c

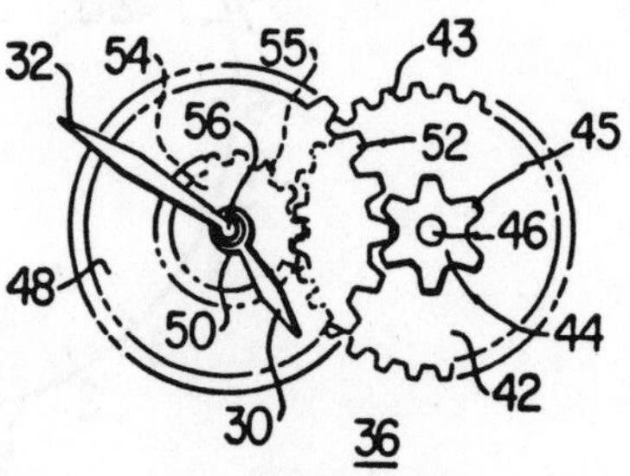

FIG. 3

FIG. 4

TWENTY-FIVE (25) HOUR CLOCK

TECHNICAL FIELD

The present invention relates generally to clock mechanisms, and more particularly to a clock for indicating twenty-five (25) simulated hours in a twenty-four (24) hour real-time day.

BACKGROUND OF THE INVENTION

A common complaint in today's fast-paced society is that there are not enough hours in the day to accomplish one's daily tasks. The cause of such complaint, however, is not of course due to an insufficient number of hours in a day; but rather one's failure to appropriately regulate his or her daily schedule or routine. This problem is exacerbated over time as one becomes conditioned to follow the same schedule or routine on a daily basis.

Scientific studies have confirmed that the human body tends to function on a twenty-five (25) hour biological system rather than the conventional twenty-four (24) hour system tied to the earth's rotation. The body's tendency to function on a twenty-five (25) hour biological clock, however, cannot be utilized advantageously since conventional timekeeping is tied to the twenty-four (24) hour hour real-time day. The twenty-five (25) hour biological clock does, however, suggest a way of ameliorating daily scheduling problems; using a twenty-five (25) simulated hour clock to provide a person with the feeling of having one extra hour per day to accomplish daily tasks.

There is therefore a need to provide a device to aid those people who are so inclined to regulate their schedules, routines and bodies to a day having twenty-five (25) simulated hours in a twenty-four (24) hour real-time day.

BRIEF SUMMARY OF THE INVENTION

Accordingly, the present invention describes a clock for indicating twenty-five (25) simulated hours in a twenty-four (24) hour real-time day. The clock preferably includes a clock face having indicia printed thereon at twenty-five (25) equally-spaced intervals, each of the intervals representing a simulated hour in a twenty-five (25) simulated hour day. The clock includes conventional hour, minute and second hands for cooperating with the clock face to provide a time indication. A clock drive mechanism drives the second hand around the clock face in 24/25th's of a real-time minute, and the minute hand around the clock face in 24/25th's of a real-time hour. Thus the hour hand is driven between spaced "hour" intervals in 1/25th of a twenty-four (24) hour real-time day.

According to the present invention, the printed indicia is appropriately selected to form either a "25 hour" clock face having the sequential numerals "1–25", or a "12—12 hour" clock face having a set of sequential numerals "1–12" on each side of the clock face and additional indicia representing the twenty-fifth (25) simulated hour of the day.

In an alternate embodiment of the invention, the printed indicia on the clock face represents a "12.5 hour" clock face. In this embodiment, the clock drive mechanism operates to drive the hour hand twice around the clock face during the twenty-four (24) hour real-time day.

BRIEF DESCRIPTION OF THE DRAWINGS

For a more complete understanding of the present invention and the advantages thereof, reference is now made to the following Description, taken in conjunction with the accompanying Drawings in which:

FIG. 1 is a perspective view of the twenty-five (25) hour clock of the present invention for indicating twenty-five (25) simulated hours in a twenty-four (24) hour real-time day;

FIGS. 2a–c show representative clock faces for use in the twenty-five (25) hour clock of FIG. 1;

FIG. 3 shows a schematic representation of a clock drive mechanism for driving the hour and minute hands of the present invention; and

FIG. 4 shows a schematic representation of a suitable digital clock drive mechanism for driving a conventional digital readout for the twenty-five (25) hour clock.

DETAILED DESCRIPTION

With reference now to the FIGURES wherein like reference characters designate like or similar parts throughout the several views, FIG. 1 is a perspective view of the twenty-five (25) hour clock 10 of the present invention. This clock indicates twenty-five (25) simulated hours in a twenty-four (24) hour real-time day. As used herein, the term "twenty-four (24) hour real-time day" refers to the twenty-four (24) hour timekeeping convention tied to the earth's rotation. The clock 10 includes a housing 12 having front and back walls 14 and 16, side walls 18, top wall 20 and a base 22. Preferably, the clock includes a clock face 24 having indicia 26 printed thereon at twenty-five (25) equally-spaced intervals 28, each of the intervals representing a simulated "hour" in a twenty-five (25) hour simulated day. The clock 10 also includes a conventional hour hand 30, minute hand 32 and second hand 34. The clock hands 30, 32 and 34 are driven by a suitable clock drive mechanism 36 for simultaneously driving the hands around the clock face 24. Although not shown in detail, the clock drive mechanism 36 is driven through a gear reduction motor 38 by a suitable power source such as an electric motor 40. Equivalent types of power sources may be used in place of the electric motor 40 as is well known in the art. The clock 10 also preferably includes a conventional clock setting mechanism (not shown) for manual setting of the hour and minute hands 30 and 32. The clock setting mechanism allows the user to reset the clock movement at any convenient time of the twenty-four (24) hour real-time day.

The twenty-five (25) hour clock 10 may be used advantageously by those people who are so inclined to regulate their schedules, routines and bodies to a day having twenty-five (25) simulated hours in a twenty-four (24) hour real-time day. The clock allows tasks to be completed in a shorter time frame and gives a person the feeling of having one extra hour per day. This extra hour, although simulated, serves to increase task efficiency by a factor of over four (4%) percent.

The theory of operation of the twenty five hour clock 10 of the present invention can be seen by considering Table I below:

TABLE I

Seconds	Minutes	Hours	Total Seconds/Day
57.6 ×	60 ×	25	= 86.400 (25 simulated hrs)

3

TABLE I-continued

Seconds	Minutes	Hours	Total Seconds/Day
60 ×	60 ×	24	= 86,400 (24 real-time hrs)

As seen in the above table, every minute in the twenty-five (25) simulated hour system includes only 24/25th's of a real-time minute, or 57.6 real-time seconds. Thus, the invention takes advantage of the saving of 2.4 real-time seconds every real-time minute to form the one extra simulated hour (57.6 real-time minutes long) every twenty-four (24) hour real-time day.

Referring simultaneously to FIGS. 2a–c, various embodiments of the clock face 24 of the clock 10 are shown. In the preferred embodiment of FIG. 2a, a "25 hour" clock face is shown wherein the indicia 26 comprise the sequential numerals "1–25" located at the twenty-five (25) equally-spaced intervals 28. Each of the intervals 28 represents a simulated "hour" in the twenty-five (25) hour simulated day, and each is separated by 14.4° (since 14.4° × 25 intervals = 360°). In FIG. 2b, a "12—12 hour" clock face is shown wherein the indicia 26 comprise sets of sequential numerals "1–12" located on each side of the clock face 24, with an additional interval 41 representing the twenty-fifth (25) simulated hour in the twenty-five (25) hour simulated day.

In FIG. 2c, the indicia 26 on the clock face 24 are located at twelve intervals, each of the intervals being 28.8° apart. An additional segment 43 of 14.4° is also used on this face to represent the extra simulated hour per day. This extra hour is allocated to one-half hour per each half day. Table II below sets forth the various timing (real-time) relationships among the clock faces 24 shown in FIGS. 2a–c.

TABLE II

	"25 Hour"	"12—12 Hour"	"12.5 Hour"
Hour Hand	1 rev./day	1 rev./day	2 rev./day
Minute Hand	1 rev./57.6 min	1 rev./57.6 min.	1 rev./57.6 min.
Second Hand	1 rev./57.6 sec	1 rev. 57.6 sec	1 rev./57.6 sec

As shown in Table II, using the "12.5 hour" face of FIG. 2c, the hour hand 30 makes two complete revolutions per twenty-four (24) hour real-time day. However, the minute hand 32 and second hand 34 move in the same fashion as with the twenty-five (25) hour clock faces of FIGS. 2a and 2b.

Referring now to FIG. 3, a suitable clock drive mechanism 36 is shown for the twenty-five (25) hour clock face of FIG. 2a. As noted above, with the twenty-five (25) hour clock face, the hour hand 30 is driven at a speed of one revolution around the clock face 24 per twenty-four (24) hour real-time day. The minute hand 32 is driven at a speed of one revolution per 24/25th's real-time minutes. The hour hand 30 is thus driven between a "hour" interval in 1/25th of the twenty-four (24) hour real-time day.

Referring to FIG. 3, the clock drive mechanism 36 includes a gear wheel 42 and pinion 44 mounted on a spindle 46 for rotation therewith. The spindle 46 is driven by a suitable power source such as the electric motor 40 as described above with respect to FIG. 1. As also shown in FIG. 3, an hour wheel 48 is mounted on a spindle 50 to drive the hour hand 30 around the clock face. The hour wheel 48 includes teeth 52 which mesh

4

with teeth 45 of the pinion 44, and thus the hour wheel 48 is driven thereby. Likewise, a minute wheel 54 is mounted on a spindle 56 to drive the minute hand 32. To this end, the minute wheel 54 includes teeth 55 which mesh with the teeth 43 of the gear wheel 42, and thus the minute wheel 54 is driven thereby.

According to the present invention, once the speed of the electric motor 38 is determined, the circumference of the hour wheel 48 and the number of teeth 52 therein are appropriately sized to drive the hour hand 30 at a speed of one revolution around the clock face 24 per 24/25th's real-time minutes. Accordingly, the hour hand 30 moves between spaced intervals 28 in 1/25th of the twenty-four (24) hour real-time day. Likewise, the circumference of the minute wheel 54 and the number of teeth 55 therein are appropriately sized to drive the minute hand 32 around the clock face 24 in 24/25th of a real-time hour. Although not shown in FIG. 3, the second hand 34 also includes a second wheel which is appropriately sized to drive the second hand 34 of FIG. 1 around the clock face 24 in 24/25th's of a real-time minute.

The clock drive mechanism 36 of FIG. 3 may also be utilized in conjunction with the clock face 24 shown in FIG. 2b since the hour hand 30 therein also makes one revolution per twenty-four (24) hour real-time day. Although not shown in detail, the clock drive mechanism 36 may also be suitably modified to drive the hour hand 30 two times around the clock face 24 in a twenty-four (24) hour real-time day. In such an embodiment, the clock face 24 shown in FIG. 2c is used.

Therefore, it can be seen that the present invention describes a unique clock mechanism for representing twenty-five (25) simulated hours in a twenty-four (24) hour real-time day. The twenty-five (25) hour clock is advantageous to those people who are so inclined to regulate their schedules, routines and bodies in such a way as to increase their efficiency by a factor of over four (4%) percent in a twenty-four (24) hour real-time day. To this end, the present invention may be used to assist a person in adjusting to a twenty-five (25) hour (57.6 real-time) minute cycle that allows tasks to be completed in a shorter time frame and provides a person with the feeling of having one extra hour per day.

Although in the preferred embodiment of the invention, the twenty-five (25) hour clock includes a clock face 24 having indicia printed thereon in twenty-five (25) equally-spaced intervals such as shown in FIGS. 2a–2b, it should be appreciated that the clock face may also comprise a digital readout representing twenty-five (25) simulated hours.

Referring now to FIG. 4, a simplified schematic diagram is shown detailing a suitable digital clock drive mechanism 60 for use in an electronic version of the twenty-five (25) hour clock. The drive mechanism 60 includes a reference oscillator 62 generating a predetermined frequency. The output of the oscillator 62 is divided by a frequency divider circuit 63 and applied to a pulse insertion/deletion circuit 64. The pulse insertion/deletion circuit 64 is suitably controlled (by a microprocessor or other conventional control circuit) to insert or delete pulses as needed in the pulse train to generate a clock signal on line 65. The clock signal is appropriately selected to generate a 57.6 second "simulated minute" and a 57.6 minute "simulated hour". This clock signal is then supplied to a conventional display drive circuit 66 which drives an LCD or LED display

5

68. The actual display readout will reset to "00:00" at the beginning of the twenty-four (24) hour real-time day and go up to 24:59" at the end of the twenty-four (24) hour real-time day. Although not shown in detail, the drive mechanism is driven by a suitable power source, such as a rechargeable battery. The electronic version of the clock also includes a conventional clock setting mechanism.

Although preferred embodiments of the invention have been described in the foregoing Detailed Description and illustrated in the accompanying Drawings, it will be understood that the invention is not limited to the embodiments disclosed, but is capable of numerous rearrangements, modifications and substitution of parts and elements without departing from the spirit of the invention. Accordingly, the present invention is intended to encompass such rearrangements, modifications and substitutions of parts and elements as fall within the spirit and scope of the appended claims.

I claim:

1. A clock, comprising:

a clock face having indicia printed thereon at twenty-five (25) equally-spaced intervals, each of said intervals representing a simulated hour in a twenty-five (25) hour simulated day;

clock hand means including an hour hand and a minute hand for cooperating with said clock face to provide a time indication; and

drive means for simultaneously driving said minute hand around said clock face and said hour hand between spaced intervals in 1/25th of said twenty-four (24) hour real-time day, whereby said clock represents twenty-five (25) simulated hours in a twenty-four (24) hour real-time day.

2. The clock for representing twenty-five (25) simulated hours as described in claim 1, wherein said clock hand means includes a second hand for cooperating with said clock face to provide a time indication.

3. The clock for representing twenty-five (25) simulated hours as described in claim 2 wherein said drive means drives said second hand around said clock face in 24/25th's of a real-time minute.

4. A clock for representing twenty-five (25) simulated hours as described in claim 1 wherein said indicia

6

printed on said clock face includes the numerals "1-25" repesenting twenty-five (25) simulated hours in said twenty-five (25) hour simulated day.

5. The clock for representing twenty-five (25) simulated hours as described in claim 1 wherein said indicia includes a set of numerals "1-12" on each side of said clock face and an additional representation of a 25th simulated hour in said twenty-five (25) hour simulated day.

6. A clock, comprising:

a clock face having indicia printed thereon at twenty-five (25) equally-spaced intervals, each of said intervals representing a simulated hour in a twenty-five (25) hour simulated day;

clock hand means including an hour hand, a minute hand, and a second hand for cooperating with said clock face to provide a time indication; and

drive means for simultaneously driving said second hand around said clock face in 24/25's of a real-time minute, said hand around said clock face in 24/25's of a real-time hour, and said hour hand between spaced intervals in 1/25th of said twenty-four (24) hour real-time day, whereby said clock represents twenty-five (25) simulated hours in a twenty-four (24) hour real-time day.

7. A clock, comprising:

a clock face having two sets of numerals "1-12" and additional indicia printed thereon at twenty-five (25) equally-spaced intervals, each of said intervals representing a simulated hour in a twenty-five (25) hour simulated day;

clock hand means including an hour hand, minute hand and second hand for cooperating with said clock face to provide a time indication; and

drive means for simultaneously driving said second hand around said clock face in 24/15's of a real-time minute, said minute hand around said clock face in 24/25th's of real-time hour, and said hour hand between spaced intervals in 1/25th of said twenty-four (24) hour real-time day, whereby said clock represents twenty-five (25) simulated hours in a twenty-four (24) hour real-time day.

* * * * *

sible to make drawings that illustrate your invention, this must be done.

Requirement F: Definition

The patent application must clearly explain what the inventor thinks his invention is. He must define exactly what it is about his invention that is different from what has been invented before.

In every patent application there is a section for "claims." In the claims section you must exactly define what your invention is. If you are claiming a chemical compound, you should state the chemical formula. If you are claiming an electric circuit, you should describe each essential element of the circuit and also describe how these fit together to get the desired result. In other words, the claims are to a pat-

ent application (and patent) what a legal description of land is to a deed.

A WORD OF ENCOURAGEMENT: WHAT IS NOT REQUIRED

To be eligible for a patent, an invention need not be a major scientific breakthrough. It does not have to produce better results than existing devices. It merely has to be (1) useful for some purpose; (2) novel, i.e., no one did it before and (3) not obvious. Your idea must be original, but you do not have to be another Thomas Edison, to say the least. The patent for a 25-Hour Clock, reproduced as Fig. 4-5, is representative of the many issued patents containing interesting ideas but not revolutionary technology. The odds are in your favor: about 70 percent of all filed patent applications ripen into patents.

CHAPTER 5

Step 1: Writing a Patent Application

The previous chapter gave you an overall understanding of what requirements a patent application must satisfy. Now you can proceed to study the four steps involved in obtaining a patent. This chapter shows you how to write the main portion of a patent application. Chapter 6 describes how to prepare other papers, such as the abstract and declaration, which are also part of the application or accompany it.

The Patent Office recommends a particular format for a patent application. This format results in a better organized application. This is the recommended format:

- ✍ Title of the Invention
- ✍ Background of the Invention
- a) Field of the Invention
- b) DESCRIPTION OF the Prior Art
- ✍ Summary of the Invention
- ✍ Brief Description of the Drawings
- ✍ Description of the Preferred Embodiments
- ✍ Claims
- ✍ Abstract of the Disclosure
- ✍ Drawings
- ✍ Oath

A sample patent application in this recommended format is shown in Fig. 5-1. I adapted the text of this sample application from U.S. Patent 4,462,599; Fig. 5-2 is the actual patent. By comparing these figures you can see how the parts of a patent application translate into the parts of an actual patent. Notice that most of the text is the same. Thus, you can learn most of what you need to know about patent application writing by simply reading issued patents. Be sure to subtitle each section, using the previous list.

TYPING AND PAPER REQUIREMENTS

It is recommended that you double space type the patent application on letter-size bond paper. Leave at least a 2-inch margin at the top of each page so that text will not be lost when patent office clerks punch holes at the top of your form during the filing process. You must include a page number at the top center of each page. It is nice to use quality legal paper with line numbers printed along the left margin. These numbers can then be referred to if you happen to discuss your patent application with the examiner over the phone, or if you need to refer to a certain part of it in written discourse.

HOW TO WRITE THE VARIOUS PARTS OF A PATENT APPLICATION

Refer to Fig. 5-1, which is a patent application typed in the recommended format. Notice the section titles and other aspects of the arrangement. All of the sub-titles and text in a patent application are imported directly into the finished patent. Thus, in the following discussion regarding how to write a patent application, we can usefully refer to publicly available patents for guidance, rather than patent applications. Here is how each of the parts of a patent application should be written.

The Title of the Invention

The title of the invention should be written at the top of page one of the application. The title is nothing more than a phrase that describes the invention. It is not critical so far as the Patent Office is concerned. The title will be published

SOCCER PRACTICE DEVICE

BACKGROUND OF THE INVENTION

1. Field of the Invention

The present Invention relates to a training device for practicing soccer skills. More specifically, the present invention relates to a training device used in teaching and practicing heading techniques.

2. Prior Art

Fear of heading a soccer ball is common among inexperienced or young players. In the past, heading has been taught by explanation and demonstration of the proper techniques followed by the teacher or coach tossing a soccer ball at a practicing player who attempts to head the ball with the top of the head.

For other sports, training devices are known which assist in development of effective techniques such as the "Jump Trainer" for basketball and volleyball practice of Alston U.S. Pat. No. 4,296,925 and the "Ball Holder" for baseball practice of Anson U.S. Pat. No. 2,772,882.

No device is known, however, for training a soccer player in effective and safe heading techniques.

SUMMARY OF THE INVENTION

The principal object of the present invention is to provide a device for use in training soccer players in effective and safe heading techniques.

It also is an object of the present invention to provide such a device which is of simple, inexpensive construction.

Fig. 5-1. Sample patent application.

Another object is to provide such a device in lightweight form that can be disassembled quickly and easily for transportation to a training site, such as in the trunk of an automobile.

A further object is to provide such a device which, in use, will decrease the fear of inexperienced players to heading a soccer ball.

The foregoing objects can be accomplished by providing a training device having a soccer ball suspended by a line from a horizontal arm cantilevered from an upright standard. In the preferred embodiment of the invention, the standard is formed by a plurality of tubular sections, the bottom section being fitted in a portable base. Each succeeding upper standard section has a bottom end portion of reduced diameter fitted in the upper end portion of the next lower section. The connection between adjacent standard sections, and the connection between the bottom standard section and the base, deter rotation of the standard relative to the base. The horizontal arm from which the soccer ball is suspended is bent horizontally outward from the top standard section, and the line suspending the soccer ball passes over a pulley carried inside the horizontal arm, then through guide eyes projecting from the arm and the standard. The height of the soccer ball can be adjusted by hauling in or paying out the line, and the line can be anchored to a cleat carried by the standard.

BRIEF DESCRIPTION OF THE DRAWINGS

Fig. 1 is a side elevation of a soccer practice device in accordance with the present invention.

Fig. 2 is a fragmentary, top perspective of the bottom end portion of the device of Fig. 1.

Fig. 3 is an enlarged, fragmentary, top perspective of a central portion of such device; and

Fig. 4 is an enlarged, fragmentary, bottom perspective of an upper portion of such device with parts broken away.

DETAILED DESCRIPTION

As shown in the drawings, the preferred soccer practice device in accordance with the present invention includes a portable base 1 of conical shape which preferable is of strong molded plastic material. The base is hollow so that it may be filled with sand, for example, through the hole normally closed by the plug 2. The upper end portion of the base forms a collar 3 into which the bottom end portion of a tubular bottom standard section 4 is fitted to maintain such section upright. A pin 5 extends through registered holes in the collar 3 and the bottom standard section 4 to prevent rotation of such section relative to the base.

The upper end portion of the bottom standard section 4 is of increased diameter for snugly receiving the lower end portion of a middle standard section 6. As best seen in Fig. 3, to prevent relative rotation of the bottom and middle standard sections an upright key plate 7 is secured to the exterior of the bottom end portion of the middle standard section and fits in an upright keyway or slot 8 in the upper end portion of the bottom standard section. Preferably, the bottom of the key plate is flush with the bottom of the middle standard section and engages against the bottom of the keyway or slot 8 so that the middle standard section does not become wedged in the bottom standard section so as to make separation of the two sections difficult.

Similarly, the upper end portion of the middle standard section 6 is of increased diameter and snugly receives the lower end portion of a top standard section 9. The connection of the top standard section to the middle standard section is the same as the connection of the middle standard section to the bottom standard section. The lower end portion of the top standard section has an upright key plate 7 secured to its exterior which plate is fitted in an upright keyway or slot 8 in the upper end portion of the middle standard section. The upper end portion of the top standard section is bent horizontally outward to form a horizontal arm 10, the free end of which is closed by a cap 11.

Preferably, all of the standard sections are lightweight metal tubing such as aluminum alloy.

A soccer ball 12 is suspended from the outer end portion of the horizontal arm 10 by a flexible line 13 which can be a rope or a cable. From the soccer ball the line extends up through a slot 14 in the underside of the outer end portion of the horizontal arm, then around a pulley or roller 15 carried by a horizontal axle 16 which can be a bolt. From the roller the cord extends down, back out through the slot 14, and then through guide eyes 17 positioned, respectively, adjacent to the inner end of the slot, at the inside of the 90 degree bend of the top standard section 9, and at the upper end portion of the middle standard section 6. The free end portion of the 13 is anchored to a cleat 18 mounted on the lower end portion of the middle standard section.

In use, the height of the soccer ball can be adjusted for an individual player by loosening the line from the cleat, hauling in or paying out the appropriate length of line and again anchoring the line to the cleat. For an inexperienced player, the ball would be positioned at about eye level and, since the ball will be stationary, the player can be instructed as to proper heading techniques and practice such techniques without fear of the ball striking the player's face, for example. As the player gains more confidence, the ball can be swung up to be headed by the player when the ball swings back down. For more experienced players, the ball can be positioned higher for practicing jumping to head the soccer ball or lower for practicing diving to head the ball.

The knockdown construction of the device allows it to be assembled and disassembled quickly and also allows it to be transported easily such as in the trunk of an automobile. Preferably, the height of the standard is at least about 11 feet (3.3 meters), in which case each standard section is about 3 to 4 feet (0.9 to 1.2 meters) long, so that the soccer ball always is suspended a substantial distance below the horizontal arm. The base can be about 2 to 3 feet (0.6 to 0.9 meter) in diameter and filled with sand to weigh about 60 to 80 pounds (27.22 to 36.29 kilograms) to support a tubular standard of a diameter of about 2 inches (50.8 millimeters) in stable fashion. The horizontal arm should be at least about 3 feet (0.9 meter) long so that a player will not contact the standard while practicing.

I claim:

1. A soccer practice device for use in practicing heading techniques comprising an elongated standard member formed of a plurality of elongated standard sections connectable end-to-end in aligned, nonrotative relationship, a portable base for receiving the lower portion of said standard member and for supporting said standard member in upright position, an elongated arm member cantilevered from the upper portion of said standard member so as to project generally horizontal when said standard member is in upright position, a soccer ball, a line having one end portion connected to said soccer ball, means carried by said members for guiding said line generally along the length of said arm member so as to suspend said soccer ball from the outer free end portion of said arm member at a height appropriate for practicing heading said soccer ball, and means carried by said standard member for attaching the free end portion of said line so as to maintain said soccer ball at such height and enabling adjustment of the height of the soccer ball.

2. A soccer practice device comprising an elongated standard member formed of a plurality of elongated tubular standard sections connectable generally end-to-end in generally aligned, nonrotative relationship, a portable base for receiving the lower portion of said standard member and for supporting said standard member in upright position, said base including connection means for deterring rotation of said standard member relative to said base, an elongated tubular arm cantilevered from the upper portion of said standard member so as to project generally horizontal when

said standard member is in upright position, said arm having a roller at generally the free end portion thereof and carried inside the bore of said arm, a soccer ball, a line having one end portion connected to said soccer ball and threaded over said roller, guide means carried by said arm for guiding said line generally along said arm, and a cleat carried by said standard member enabling the free end portion of said line to be anchored to said cleat for maintaining said soccer ball at a desired height selected for an individual player and appropriate for such player to practice heading the soccer ball.

3. A soccer practice device comprising a long and straight tubular bottom standard section, a long and straight tubular middle standard section of approximately the same length as said bottom standard section, a tubular top standard section having a first long and straight portion of approximately the same length as said bottom and middle standard sections and a second long and straight portion bent generally perpendicular to said first portion and of approximately the same length as said first portion, standard connection means for joining said standard sections generally end-to-end with said bottom standard section, said middle standard section and said first portion of said top standard section in generally coaxial relationship, said standard connection means including means for deterring relative rotation of said standard sections, a portable base, base-connection means for joining an end portion of said bottom standard section to said portable base and including means for deterring relative rotation of said bottom standard section and said base, a soccer ball, a

line having one end portion connected to said soccer ball, means for guiding said line generally along said arm from generally the free end portion of said arm to generally the area of connection of said first and second portions of said arm, and anchor means carried by one of said standard sections for anchoring thereto the free end portion of said line.

ABSTRACT OF THE DISCLOSURE

A soccer ball is suspended by a line from a horizontal arm cantilevered from an upright standard supported by a portable base. The height of the soccer ball can be adjusted by paying out or hauling in the appropriate length of line, and the line can be anchored to a cleat carried by the standard. The height of the ball is adjusted for soccer player for instruction and practice of heading the ball. The standard is of knockdown construction for case in transportation to a training site.

in various indexes and summaries of patents issued, for example, in the *Official Gazette* of the patent office. Because these summaries are used by people who are looking for particular types of technology, you should write your title in a way that will attract them. For example, if you have found that methyl choloride cures warts, your title will more likely attract people looking for such a thing if you make it "USE OF METHYL CHLORIDE AS WART CURE" rather than "USE FOR METHYL CHOLORIDE." The latter version of the title tells people nothing about the economic value of your invention. Don't go too far, however. The Patent Office will not let you get away with laudatory titles, such as "TREMENDOUS WART CURE." They are in the business of printing patents, not advertising circulars. The title must be brief but technically accurate, preferably in the range of two to seven words.

The Background of the Invention

The Background of the Invention section has two parts: (1) The field of the invention and (2) description of related art and prior art.

THE FIELD OF THE INVENTION

The field of the invention section should have just one or two sentences. The purpose of this sentence is to help Patent Office clerks forward the patent application to the proper patent examiner and also assign it the proper technical category numbers.

An acceptable single sentence format for the Field of the Invention would be: "This invention relates in general to _____ and in particular to _____." For Example,

"This invention relates in general to computers, and in particular to computers with numerous arithmetic processors operating in parallel."

"This invention relates in general to artificial lighting devices, and in particular to lighting devices wherein light is produced by causing current to flow through a gas."

"This invention relates in general to insecticide compositions, and in particular to a roach-killing composition comprising aspartame and powdered portland cement."

THE DESCRIPTION OF RELATED ART

In the Description of Related Art section you begin to tell the story that is behind the invention. That's right—a good patent application should contain a story. A typical plot for the story in your patent application has these parts:

- There is a problem to be solved.
- Other people have tried to solve this problem in various ways. (Describe the various ways.)
- These people have not completely solved the problem. (Describe why the solutions proposed previously are inadequate.)

- But I, Joe Inventor, have discovered the solution!
- Here is how my remarkable invention works. (Describe details.)
- As a reward for my contribution, I claim rights to the following invention. (Define invention.)

Parts 1, 2, and 3 of the plot go into the Description of the Related Art section of the patent application. Parts 4, 5, and 6 are covered in the summary of the invention, the detailed description, and the claims sections. If a patent has a "plot," as described above, it will be better appreciated by future readers, such a federal judges and juries. The plot will lead these readers to conclude that you, the hero, have done something deserving of financial reward.

The purpose of the Related Art section is to describe previous technology and explain the shortcomings of this technology. It should introduce the reader to the technology, explain what the technology is supposed to accomplish, and explain how previous patents and inventions described in the technical literature have attempted to achieve the goals. You should give the numbers of the previous patents and volume and page citations to other forms of literature.

The "Background of the Invention Section" of Fig. 5-1 has two parts. Part l, entitled Field of the Invention Section, has been described. Part 2 is entitled "Prior Art." If you wish, you could say "Description of Related Art," instead of "Prior Art."

The invention shown in Fig. 5-1 is quite simple and therefore does not need an elaborate Background of the Invention section. Fig. 5-3 shows a more complicated electronic device patent owned by IBM. This patent has an elaborate, well-written background section which begins with the following language:

"In order to provide background information so that the invention may be completely understood and appreciated in its proper context, reference is made to a number of prior art patents and publications as follows:"

Likewise, at column 2, line 4 there is this language:

"Whatever the precise merits, features and advantages of the above cited references, none of them achieves or fulfills the purposes of . . . the present invention."

This is typical boilerplate language. It lets the reader know which part of the plot comes next.

As you read the background, notice that many prior patents and articles are referenced. In general, a separate paragraph is devoted to each reference. The first sentence in this paragraph describes the previous technology. The last sentence in the paragraph states how the old technology differs from the present invention. This helps future readers, such as Federal judges and juries, better appreciate the contribution made by the inventor and feel more inclined to protect his efforts. This discussion also puts the patent examiner on the right track when he does his search for prior art. If you explain to the examiner in your background section why these

United States Patent [19]

Brown

[11] **Patent Number:** **4,462,599**

[45] **Date of Patent:** **Jul. 31, 1984**

[54] **SOCCER PRACTICE DEVICE**

[76] Inventor: **Ralph Brown,** c/o Melvin Clark, 4234 Interlake Ave. N., Seattle, Wash. 98103

[21] Appl. No.: **537,998**

[22] Filed: **Sep. 30, 1983**

[51] Int. Cl.³ A63B 69/00
[52] U.S. Cl. .... 273/411; 273/58 C; 273/413
[58] Field of Search .... 273/413, 411, 58 C, 273/26 E, 26 EA, 29 A

[56] **References Cited**

U.S. PATENT DOCUMENTS

2,772,882	12/1956	Anson	....	273/26 E
3,262,703	7/1966	Hodlick	....	273/413
4,158,458	6/1979	Gomez	....	273/413
4,191,372	3/1980	Keller	....	273/29 A
4,296,925	10/1981	Alston	....	273/413 X

FOREIGN PATENT DOCUMENTS

2508922 9/1976 Fed. Rep. of Germany 273/411

Primary Examiner—William H. Grieb
Attorney, Agent, or Firm—Ward Brown; Robert W. Beach

[57] **ABSTRACT**

A soccer ball is suspended by a line from a horizontal arm cantilevered from an upright standard supported by a portable base. The height of the soccer ball can be adjusted by paying out or hauling in the appropriate length of line, and the line can be anchored to a cleat carried by the standard. The height of the ball is adjusted for a soccer player for instruction and practice of heading the ball. The standard is of knockdown construction for ease in transportation to a training site.

3 Claims, 4 Drawing Figures

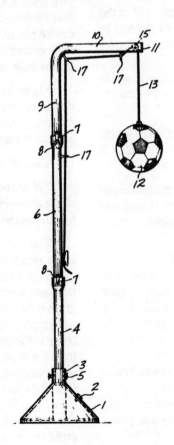

Fig. 5-2. Patent resulting from the sample application.

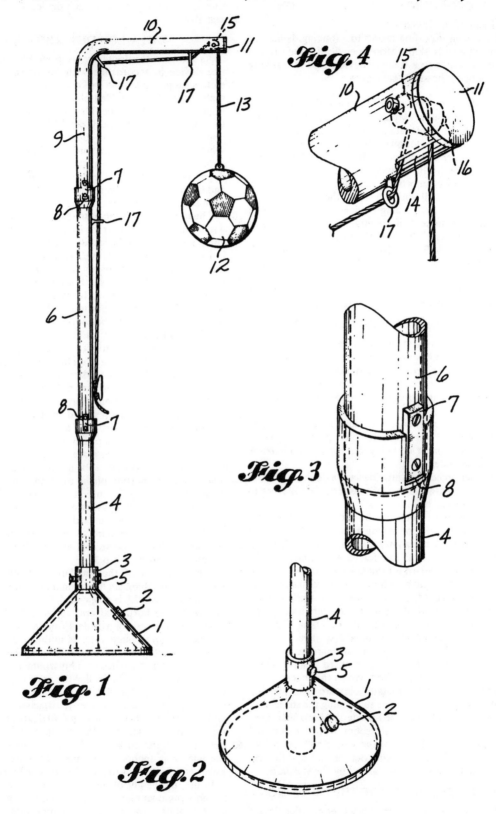

Fig.4

Fig.3

Fig.1

Fig.2

1

SOCCER PRACTICE DEVICE

BACKGROUND OF THE INVENTION

1. Field of the Invention

The present invention relates to a training device for practicing soccer skills. More specifically, the present invention relates to a training device used in teaching and practicing heading techniques.

2. Prior Art

Fear of heading a soccer ball is common among inexperienced or young players. In the past, heading has been taught by explanation and demonstration of the proper techniques, followed by the teacher or coach tossing a soccer ball at a practicing player who attempts to head the ball. This method can increase the player's fear and reinforce improper techniques such as attempting to head the ball with the top of the head.

For other sports, training devices are known which assist in development of effective techniques, such as the "Jump Trainer" for basketball and volleyball practice of Alston U.S. Pat. No. 4,296,925 and the "Ball Holder" for baseball practice of Anson U.S. Pat. No. 2,772,882.

No device is known, however, for training a soccer player in effective and safe heading techniques.

SUMMARY OF THE INVENTION

The principal object of the present invention is to provide a device for use in training soccer players in effective and safe heading techniques.

It also is an object of the present invention to provide such a device which is of simple, inexpensive construction.

Another object is to provide such a device in lightweight form that can be disassembled quickly and easily for transportation to a training site, such as in the trunk of an automobile.

A further object is to provide such a device which, in use, will decrease the fear of inexperienced players to heading a soccer ball.

The foregoing objects can be accomplished by providing a training device having a soccer ball suspended by a line from a horizontal arm cantilevered from an upright standard. In the preferred embodiment of the invention, the standard is formed by a plurality of tubular sections, the bottom section being fitted in a portable base. Each succeeding upper standard section has a bottom end portion of reduced diameter fitted in the upper end portion of the next lower section. The connection between adjacent standard sections, and the connection between the bottom standard section and the base, deter rotation of the standard relative to the base. The horizontal arm from which the soccer ball is suspended is bent horizontally outward from the top standard section, and the line suspending the soccer ball passes over a pulley carried inside the horizontal arm, then through guide eyes projecting from the arm and the standard. The height of the soccer ball can be adjusted by hauling in or paying out the line, and the line can be anchored to a cleat carried by the standard.

BRIEF DESCRIPTION OF THE DRAWINGS

FIG. 1 is a side elevation of a soccer practice device in accordance with the present invention.

FIG. 2 is a fragmentary, top perspective of the bottom end portion of the device of FIG. 1;

2

FIG. 3 is an enlarged, fragmentary, top perspective of a central portion of such device; and

FIG. 4 is an enlarged, fragmentary, bottom perspective of an upper portion of such device with parts broken away.

DETAILED DESCRIPTION

As shown in the drawings, the preferred soccer practice device in accordance with the present invention includes a portable base 1 of conical shape which preferably is of strong molded plastic material. The base is hollow so that it may be filled with sand, for example, through the hole normally closed by the plug 2. The upper end portion of the base forms a collar 3 into which the bottom end portion of a tubular bottom standard section 4 is fitted to maintain such section upright. A pin 5 extends through registered holes in the collar 3 and the bottom standard section 4 to prevent rotation of such section relative to the base.

The upper end portion of the bottom standard section 4 is of increased diameter for snugly receiving the lower end portion of a middle standard section 6. As best seen in FIG. 3, to prevent relative rotation of the bottom and middle standard sections an upright key plate 7 is secured to the exterior of the bottom end portion of the middle standard section and fits in an upright keyway or slot 8 in the upper end portion of the bottom standard section. Preferably, the bottom of the key plate is flush with the bottom of the middle standard section and engages against the bottom of the keyway or slot 8 so that the middle standard section does not become wedged in the bottom standard section so as to make separation of the two sections difficult.

Similarly, the upper end portion of the middle standard section 6 is of increased diameter and snugly receives the lower end portion of a top standard section 9. The connection of the top standard section to the middle standard section is the same as the connection of the middle standard section to the bottom standard section. The lower end portion of the top standard section has an upright key plate 7 secured to its exterior which plate is fitted in an upright keyway or slot 8 in the upper end portion of the middle standard section. The upper end portion of the top standard section is bent horizontally outward to form a horizontal arm 10, the free end of which is closed by a cap 11.

Preferably, all of the standard sections are lightweight metal tubing such as aluminum alloy.

A soccer ball 12 is suspended from the outer end portion of the horizontal arm 10 by a flexible line 13 which can be a rope or a cable. From the soccer ball the line extends up through a slot 14 in the underside of the outer end portion of the horizontal arm, then around a pulley or roller 15 carried by a horizontal axle 16 which can be a bolt. From the roller the cord extends down, back out through the slot 14, and then through guide eyes 17 positioned, respectively, adjacent to the inner end of the slot, at the inside of the 90 degree bend of the top standard section 9, and at the upper end portion of the middle standard section 6. The free end portion of the line 13 is anchored to a cleat 18 mounted on the lower end portion of the middle standard section.

In use, the height of the soccer ball can be adjusted for an individual player by loosening the line from the cleat, hauling in or paying out the appropriate length of line and again anchoring the line to the cleat. For an inexperienced player, the ball would be positioned at about eye level and, since the ball will be stationary, the

3

player can be instructed as to proper heading techniques and practice such techniques without fear of the ball striking the player's face, for example. As the player gains more confidence, the ball can be swung up to be headed by the player when the ball swings back down. For more experienced players, the ball can be positioned higher for practicing jumping to head the soccer ball or lower for practicing diving to head the ball.

The knockdown construction of the device allows it to be assembled and disassembled quickly and also allows it to be transported easily such as in the trunk of an automobile. Preferably, the height of the standard is at least about 11 feet (3.3 meters), in which case each standard section is about 3 to 4 feet (0.9 to 1.2 meters) long, so that the soccer ball always is suspended a substantial distance below the horizontal arm. The base can be about 2 to 3 feet (0.6 to 0.9 meter) in diameter and filled with sand to weigh about 60 to 80 pounds (27.22 to 36.29 kilograms) to support a tubular standard of a diameter of about 2 inches (50.8 millimeters) in stable fashion. The horizontal arm should be at least about 3 feet (0.9 meter) long so that a player will not contact the standard while practicing.

I claim:

1. A soccer practice device for use in practicing heading techniques comprising an elongated standard member formed of a plurality of elongated standard sections connectible end-to-end in aligned, nonrotative relationship, a portable base for receiving the lower portion of said standard member and for supporting said standard member in upright position, an elongated arm member cantilevered from the upper portion of said standard member so as to project generally horizontal when said standard member is in upright position, a soccer ball, a line having one end portion connected to said soccer ball, means carried by said members for guiding said line generally along the length of said arm member so as to suspend said soccer ball from the outer free end portion of said arm member at a height appropriate for practicing heading said soccer ball, and means carried by said standard member for attaching the free end portion of said line so as to maintain said soccer ball at such height and enabling adjustment of the height of the soccer ball.

2. A soccer practice device comprising an elongated standard member formed of a plurality of elongated tubular standard sections connectible generally end-to-

4

end in generally aligned, nonrotative relationship, a portable base for receiving the lower portion of said standard member and for supporting said standard member in upright position, said base including connection means for deterring rotation of said standard member relative to said base, an elongated tubular arm cantilevered from the upper portion of said standard member so as to project generally horizontal when said standard member is in upright position, said arm having a roller at generally the free end portion thereof and carried inside the bore of said arm, a soccer ball, a line having one end portion connected to said soccer ball and threaded over said roller, guide means carried by said arm for guiding said line generally along said arm, and a cleat carried by said standard member enabling the free end portion of said line to be anchored to said cleat for maintaining said soccer ball at a desired height selected for an individual player and appropriate for such player to practice heading the soccer ball.

3. A soccer practice device comprising a long and straight tubular bottom standard section, a long and straight tubular middle standard section of approximately the same length as said bottom standard section, a tubular top standard section having a first long and straight portion of approximately the same length as said bottom and middle standard sections and a second long and straight portion bent generally perpendicular to said first portion and of approximately the same length as said first portion, standard connection means for joining said standard sections generally end-to-end with said bottom standard section, said middle standard section and said first portion of said top standard section in generally coaxial relationship, said standard connection means including means for deterring relative rotation of said standard sections, a portable base, base-connection means for joining an end portion of said bottom standard section to said portable base and including means for deterring relative rotation of said bottom standard section and said base, a soccer ball, a line having one end portion connected to said soccer ball, means for guiding said line generally along said arm from generally the free end portion of said arm to generally the area of connection of said first and second portions of said arm, and anchor means carried by one of said standard sections for anchoring thereto the free end portion of said line.

* * * * *

United States Patent [19]

Dansky et al.

[11] Patent Number: **4,605,870**

[45] Date of Patent: **Aug. 12, 1986**

[54] **HIGH SPEED LOW POWER CURRENT CONTROLLED GATE CIRCUIT**

[75] Inventors: **Allan H. Dansky,** Poughkeepsie; **John P. Norsworthy,** Fishkill, both of N.Y.

[73] Assignee: **IBM Corporation,** Hopewell Junction, N.Y.

[21] Appl. No.: **478,613**

[22] Filed: **Mar. 25, 1983**

[51] Int. Cl.⁴ H03K 3/33; H03K 17/04; H03K 19/013; H03K 19/088

[52] U.S. Cl. 307/443; 307/456; 307/300; 307/280

[58] Field of Search 307/454, 455, 456, 443, 307/467, 475, 457, 458, 270, 280, 300, 319, 320

[56] **References Cited**

U.S. PATENT DOCUMENTS

2,964,652	12/1960	Yourke	307/88.5
3,505,535	4/1970	Cavaliere	307/203
3,614,467	10/1971	Tu	307/300 X
3,676,708	7/1972	Uchida	307/319 X
3,909,637	9/1975	Dorler	307/455 X
4,092,551	5/1978	Howard et al.	307/254
4,132,906	1/1979	Allen	307/443
4,306,159	12/1981	Wiedmann	307/457
4,321,485	3/1982	Morozowich et al.	307/300
4,330,723	5/1982	Griffith	307/458 X
4,453,089	6/1984	Shuey et al.	307/280 X

OTHER PUBLICATIONS

IBM Technical Disclosure Bulletin, Dansky et al., Current-Controlled Gate Push–Pull Dotting, Nov. 1981, pp. 3031–3034.

IBM Technical Disclosure Bulletin, Barish et al., Current Switch Push–Pull Internal Circuit, Nov. 1981, p. 3041.

Primary Examiner—Stanley D. Miller
Assistant Examiner—David R. Bertelson
Attorney, Agent, or Firm—John F. Ohlandt

[57] **ABSTRACT**

The invention pertains to semiconductor circuitry, and more particularly to a class of circuitry known as current controlled gate circuits for driving very large scale integrated circuit gate arrays; the novel circuit can achieve much lower speed-power products than other circuitry, such as the well known T²L circuitry; the circuit includes push-pull drive and it provides negligible DC current in both DC states, that is, On and Off.

8 Claims, 3 Drawing Figures

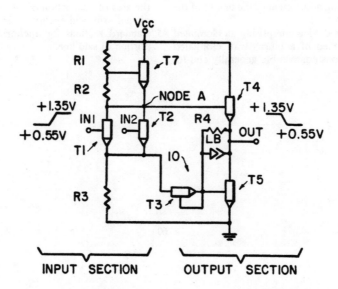

Fig. 5-3. Patent: High Speed, Low Power Current Controlled Gate Circuit.

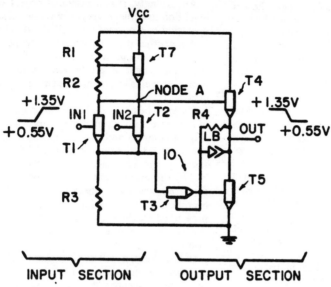

FIG.IA

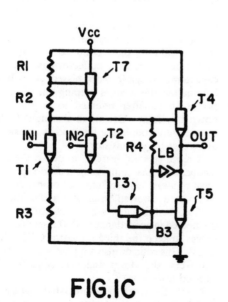

FIG.IC

FIG.IB

HIGH SPEED LOW POWER CURRENT CONTROLLED GATE CIRCUIT

BACKGROUND OF THE INVENTION

1. Field of the Invention

This invention pertains to semiconductor digital circuits, and more particularly, to a branch or field thereof known as current controlled gates, which are bipolar logic circuits intended for use on dense, very large scale integrated circuit gate arrays. In this particular field of current controlled gate circuits, the emphasis is upon extending the art to ever greater densities and to the achievement of lower propagation delays.

2. Background Information

In order to provide background information so that the invention may be completely understood and appreciated in its proper context, reference may be made to a number of prior art patents and publications as follows:

U.S. Pat. Nos. 2,964,652; 3,505,535; and 4,092,551; all of which are assigned to the assignee of the present invention; Article in IBM Technical Disclosure Bulletin Vol. 24 No. 6, pages 3031–3034, November 1981, entitled "Current Controlled Gate Push-Pull Dotting"; U.S. patent application Ser. No. 221,684 filed Dec. 30, 1980, and assigned to the assignee of the present invention.

U.S. Pat. No. 3,505,535 to Cavaliere in particular discloses an improvement on what has come to be known in the art as the "current switch", a form of circuitry first disclosed in U.S. Pat. No. 2,964,652 to H. S. Yourke. In the current switch a constant current is switched either to one or more logic input transistors, or to a grounded-base transistor, depending upon potential levels of the logic signals at the bases of the input transistors in relation to the reference potential at the grounded base transistor. Because the current which flows through the collector load resistors is constant and predetermined, circuit parameters may be selected so as to limit the potential swing of the collectors, thereby to maintain the transistors out of saturation. Nevertheless, despite the useful applications of the current switch, per se, U.S. Pat. No. 3,505,535 provides an improved clamp for the basic circuit, thereby to insure that the transistors therein are prevented from going deeply into the saturation region. This is achieved by a non-linear load impedance network in the form of a transistor at the output of the logic input transistors.

U.S. Pat. No. 4,092,551 to Howard et al discloses a so-called "speed-up" circuit, which uses a base-collector junction as a capacitor in order to aid in discharging the base of a saturated transistor; however, the present invention distinguishes therefrom as will be made apparent in the later description.

The article cited which appeared in the IBM Technical Disclosure Bulletin for November 1981 discloses a current controlled gate circuit which includes a Schottky barrier diode, connected between the emitter of one output transistor and the collector of the other. However, that circuit is directed to solving the difficult problem associated with the "dotting" of push-pull outputs, and the form and function of the circuit is similar to U.S. patent application Ser. No. 221,684 cited above.

Another disclosure which furnishes background information is that above cited U.S. patent application Ser. No. 221,684 in the name of J. A. Dorler et al, which discloses a transient controlled current switch in several forms or embodiments. However, that circuit does not operate in the manner provided by the present invention.

Whatever the precise merits, features and advantages of the above cited references, none of them achieves or fulfills the purposes of the current controlled gate circuit of the present invention.

Accordingly, it is a principal object of the present invention to achieve very low power dissipation in the operation of a current controlled gate circuit.

It is another principal object of the present invention to provide a circuit with characteristics that will enable the achievement of greater density in large scale gate arrays, as well as the achievement of lower propagation delays.

It should be noted that in the description of the circuit of the present invention it may appear that the topology is quite complex and, therefore, this might seem inconsistent with the circuit being used for a dense "master slice". However, density in gate arrays is actually limited by the space required to interconnect the gates and therefore, the space that the silicon devices occupy becomes irrelevant as long as they fit into a rectangle prescribed by the required wiring channels per gate in the two levels above the gate dedicated to interconnection metal.

In any event, it will be clear that to advance the art it is necessary that a logic circuit have minimal propagation delay. This propagation delay may be thought of as being composed of two parts: the unloaded delay, and the added delay due to loading (e.g. fan out, wiring capacitance, etc.). The delay per unit load is called the sensitivity due to loading (for example: 0.1 ns/pf or 0.3 ns/fo, where ns is nanoseconds, pf is picofarads and fo is fan out).

SUMMARY OF THE INVENTION

In fulfillment and implementation of the previously recited objects, a primary feature of the invention resides in the provision of a unique coupling arrangement for coupling the input section to a push-pull output section in the current controlled gate circuit. Included in this coupling arrangement is a transistor having its emitter-and-base short circuited so that that particular transistor serves as a collector-base diode with high capacitance when forward biased. As will be explained hereinafter, this enables keeping one of the output transistors—hereinafter referred to as the third or pull-down transistor—ON when the input is down and to capacitively couple the input so as transiently to overdrive that third transistor.

A more specific feature resides in the inclusion of a feedback means comprising a diode, preferably in the form of a low barrier Schottky diode, connected to the aforenoted "shorted" transistor and to the output of the circuit. Additionally, a base resistor which is variably connected in accordance with the several embodiments, serves to keep the third transistor OFF when the input is UP. This resistor also supplies a small current to forward bias the aforenoted base-collector diode of the shorted transistor.

It will, therefore, be appreciated that by virtue of the coupling arrangement provided, the output section of the controlled gate circuit is active only during an output transition, consuming negligible power in either of the two DC states of the circuit. Moreover, a low supply voltage of approximately 2.1 volts to ground is used.

Furthermore, since the input section is buffered from loading factors at the output by the push-pull output section, resistances can be made large without causing unacceptably long delays. By increasing the resistances involved, power is further conserved.

It will also be understood from the detailed description which follows that the circuit of the present invention exploits inherent device capacitances to accelerate the downgoing transition of the output voltage of the gate. In conjunction with the "pull-up" provided by the emitter follower (second transistor) of the push-pull output section, lower delays are realized.

Other and further objects, advantages and features of the present invention will be understood by reference to the following specification in conjunction with the annexed drawing, wherein like parts have been given like numbers.

BRIEF DESCRIPTION OF THE DRAWING

FIG. 1A is the most preferred embodiment of the current controlled gate circuit of the present invention.

FIGS. 1B and 1C are schematic diagrams of alternate embodiments of the circuit of the present invention.

DESCRIPTION OF PREFERRED EMBODIMENTS

Before proceeding with a detailed description of the present invention, it should be noted that a description of several embodiments of the invention has already appeared in the IBM Technical Disclosure Bulletin, Vol. 24, No. 11A, and pages 5613–5618, April 1982. Accordingly, that article is incorporated herein by reference. The voltage wave forms contained therein may be consulted for an appreciation of the operation of the circuitry to be described.

Referring now specifically to FIG. 1A, there is illustrated the most preferred embodiment of the circuit inasmuch as it dissipates minimal power. It will be understood that although several input transistors, that is, T1 and T2, have been shown in the circuit to perform a NOR function, only a single transistor need be used in the input section for operation as an inverter. The input section also includes transistor T7 which functions as a clamp to prevent devices T1 and T2 from saturating at high temperature and high supply.

A voltage source Vcc is connected to both the input and output sections and a voltage divider, comprising R1 and R2, is provided for suitably biasing transistor T7. An output resistor R3 is connected in common to the emitters of transistors T1 and T2. Transistors T4 and T5 are included in the push-pull output section, the transistor T4 functioning basically as an emitter follower, while T5 functions as an AC coupled active pull-down device.

The unique coupling arrangement of the present invention is designated 10. This means for coupling extends from the emitter resistor R3 to the base input of transistor T5. The coupling means includes the transistor T3, which, as seen, has its emitter and base short circuited. Accordingly, this transistor functions as a base-to-collector diode which provides high capacitance when the input is initially Down. Furthermore, this capacitive element in the form of the base-to-collector diode enables capacitive coupling of the input signal to the base of transistor T5, thereby to drive T5 very hard ("overdrive") in the transient state, thereby giving a fast down-going transition at the output of the gate circuit.

As noted previously, the base resistor R4, seen in this particular embodiment to be connected from the output to the input of transistor T5, provides dc feedback to turn Off transistor T5 completely when the output of the circuit reaches a Down level. In this case, the current in T5 is in the microampere range. Resistor R4 also functions when the output is Up, to supply a small current to forward bias the base-collector diode of transistor T3. Consequently, R4 forms, with such base-collector diode, a biasing network to maintain the active pull down transistor T5 in a low current state, which is somewhat higher than the current state when the output is Down, but still is in the microampere range.

The low barrier Schottky diode LB is connected as a feedback means from the output of transistor T5 to its input, and gives an output voltage undershoot to about 0.5 volts, causing the Schottky diode LB to turn itself On, thereby turning Off transistor T5. The term "output voltage undershoot" refers to the fact that the downgoing wave or pulse dips further, for a brief period, than the level it obtains in its more permanent state. Thus, this device functions to provide dynamic feedback from the output of the circuit to the base of transistor T5 in order to reduce the high transient current in T5 to a low DC current when the output voltage state has switched and attahced a suitable Down level.

Operation

In describing the circuit operation, reference may be made to FIG. 1A in which the most preferred embodiment of the current controlled gate circuit is illustrated. The currents and voltages in the circuit are specified in accordance with the following:

IR3 . . . current in resistor R3
IB1 . . . base current in transistor T1
IE3 . . . emitter current in transistor T3
IE4 . . . emitter current in transistor T4
IC3 . . . collector current in transistor T3
VBE1 . . . base to emitter voltage of transistor T1
VBE2 . . . base to emitter voltage of transistor T2
VBC3 . . . base to collector voltage of transistor T3
VBE5 . . . base to emitter voltage of transistor T5
VA . . . voltage with respect to ground at node A
V_{in} . . . input voltage of the gate circuit.

It will be understood that there are two DC states of the current controlled gate circuit, one being when all of the inputs are Down, and the other being when any of the inputs are Up. When the input IN1 is Up, for example, T1 conducts a current approximately equal to $(V_{in} - VBE1)/R3$. This current is also carried by the anti-saturation clamp network described in U.S. Pat. No. 3,505,535 previously cited. This network consists of R1, R2 and T7. As a consequence, node A drops to about 1.1 volts. At this point, transistors T3, T4, and T5 are all operating in an extremely low current region (about 0.6 to 10 microamperes). Transistors T4 and T5 are each marginally On, thereby resulting in an output level of about 1.1 v/2=0.55 v., as indicated on FIG. 1A. A trickle of current passes through R4 and into the base of T5, thereby producing the marginally On state of T5. The output level of about 0.55 volts, which corresponds to the input being at an Up level of approximately 1.35 volts, is the Down level of the output.

On the other hand, when the inputs of transistors T1 and T2 are Down, that is, about 0.55 volts as indicated on FIG. 1A, transistors T1 and T2 are Off and consequently, VA rises to about V_{CC}. Therefore, transistor T4 becomes conducting so that $V_{out}=V_{cc}-VBE4$, which is approximately equal to 1.35 volts. This voltage corresponds to the Up level of the circuit. IE4 flows through R4, across the base-collector junction of T3, through R3, into ground. Resistor R4 is made much larger than resistor R3 so that almost all of the voltage drop due to this current appears across resistor R4. The small voltage drop across R3 plus the collector-to-base junction of transistor T3 form a "current mirror" with transistor T5. Due to the small voltage drop across R3, T3 should be designed with a large base-to-collector area to keep the current in T5 within an order of magnitude of the current in the R4-T3 combination. This current in the base-collector junction of transistor T3 acts to increase the capacitance of that junction, which will be exploited during the Down-going transition of the output voltage.

As indicated previously, the push-pull output section of the circuit of FIG. 1A provides for active transition in both directions, thereby providing high speed. When the input IN1 falls, transistor T1 turns Off and VA rises to the power supply level V_{cc}. The output follows node A Up since transistor T4 acts as an emitter follower. The circuit then rests in the DC state with the input Down, and the output UP.

In the other active transition, that is, when the input rises, transistor T3 acts as a capacitor to connect the emitter of transistor T1 to the base of transistor T5. The base-collector junction of T3 had been forward biased and, due to the presence of resistor R4, was carrying a small amount of current immediately prior to switching, which means that a fairly large capacitance existed across that junction at that time This causes VBE5 to rise quickly, thereby turning ON transistor T5, thereby creating a low impedance between the output of the circuit and ground. Thus the output voltage falls quite quickly. At the same time, the current conducted in transistor T1 passes through the antisaturation clamping network, causing node A to drop to about 1.1 volts; consequently, emitter follower T4 is nearly turned Off. Diode LB is a low barrier Schottky diode. When the output falls to a point such that VBC5 is about 0.3 volts, diode LB starts to conduct and drains excess charge from T3 through the collector of transistor T5, thereby causing the circuit to stabilize.

Although the circuit of FIG. 1A is considered the most preferred embodiment because it has been found to dissipate a minimum of power, other alternate embodiments are possible. In particular, these embodiments envision different connections for the resistor R4. Thus, as seen in FIG. 1B, instead of this resistor being connected from the emitter of T4 to the base of T5, it is connected between the base of T5 and the power supply V_{cc}, (thereby providing a current sink when driving a different type of load circuit from the one illustrated).

Another further embodiment of what is essentially the same circuit as seen in FIG. 1A involves the connection of R4 to the base of transistor T4, rather than to the emitter thereof as in FIG. 1A. FIG. 1C shows this embodiment, in which the resistor R4 can be made smaller than is the case in FIG. 1B.

To assist the man skilled in the art in the practice of the circuit of the present invention, a somewhat detailed specification is offered herewith:

All of the transistors for example in FIG. 1A, are NPN switching transistors; the resistor R3 has a value of approximately 1.0 kilohms, the resistor R4 has a value of 15 kilohms. The voltage V_{cc} as indicated previously is 2.1 volts approximately. Accordingly, when the circuit of FIG. 1A with the values indicated is on the On state, such that at least one of the inputs has an Up value of 1.35 volts, (the output being Down), then a very small current flows through the push-pull output section; that is, through transistors T4 and T5, such current having a value of approximately 0.6 microamperes. The input current in the On condition is of the order of 500 to 1000 times greater than the output current value of 0.6 microamperes. Thus, it will be appreciated that, although the value of the output current may be varied in accordance with modifications of the circuit, the range of ratios of input to output current should be maintained in all cases.

The current controlled gate circuits in accordance with the present invention represent a means for achieving improved performance and lower power dissipation. The characteristic low sensitivity to capacitive loading makes it ideal for gate array applications, since delays can be well predicted before any circuits are automatically wired. If, for example, a logic gate in the critical path of a synchronous computing machine had excessive capacitive loading at its output after automatic placement of wiring, the desired machine cycle would not suffer greatly.

While there have been shown and described what are considered at present to be the preferred embodiments of the present invention, it will be appreciated by those skilled in the art that modifications of such embodiments may be made. It is therefore desired that the invention not be limited to these embodiments, and it is intended to cover in the appended claims all such modifications as fall within the true spirit and scope of the invention.

We claim:

1. A high speed, low power, current-controlled logic gate circuit comprising:

a power supply;

a first section including at least a first transistor having a logic input, which corresponds to the input of said gate circuit, and first and second outputs;

a second, push-pull, section having an output, which corresponds to the output of said gate circuit, said second section further including at least interconnected second and third transistors; the second transistor having an input connected to the first output of said first section; the third transistor having an input;

means for coupling the second output of said first section directly to the input of said third transistor, comprising means, including an asymmetrically conductive device and the inherent capacitance of said device, for overdriving and turning ON said third transistor during switching of said gate circuit from a state in which the output is at an UP level to a state in which the output is at a DOWN level;

a resistor, connected to said asymmetrically conductive device so as to form with said device a biasing network for maintaining the third, or active pull-down, transistor in a low current ON state, which is in the microampere range, when the output of said gate circuit is at said UP level, said resistor also being operable at that time for supplying current to forward bias said asymmetrically conductive device, said resistor further providing DC feedback to turn OFF said third transistor when the output of said gate circuit reaches said Down level such that a very low current then flows through said third transistor, said very low current having a

4,605,870

7

value of approximately 1 microampere, which is serveral times lower than said low current;

a Schottky diode connected as a dynamic feedback means from the output of said gate circuit to the input of said third transistor, said Schottky diode giving an output voltage undershoot, thereby turning itself ON and turning said third transistor OFF responsive to the output of said gate circuit switching to its DOWN level.

2. A circuit as defined in claim 1 in which said very low current at the output, when the output is Down, has a value of approximately 0.6 microamperes.

3. A circuit as defined in claim 1, in which the ratio between the input current and the output current, when the output is Down, is 500 to 1,000.

4. A circuit as defined in claim 1, in which said asymmetrically conductive device comprises a fourth transistor, said inherent capacitance being defined by the col-

8

lector-base junction of said fourth transistor which has its base and emitter short circuited.

5. A circuit as defined in claim 4, in which said resistor is connected between the emitter of said fourth transistor and the emitter of said second transistor.

6. A circuit as defined in claim 4, in which said resistor is connected between the emitter of said fourth transistor and the base of said second transistor.

7. A circuit as defined in claim 4, in which said resistor is connected between the emitter of said fourth transistor and the power supply.

8. A circuit as defined in claim 4, further comprising an emitter resistor for said first transistor, the collector of said collector-base junction of said fourth transistor being connected to said emitter resistor, the base of said fourth transistor being connected to the base of said third transistor.

* * * * *

references do not do the same job as does your invention, he is less likely to send out a rejection when he discovers the references for himself in the patent office search room.

You should discuss every close reference that you know about. By all means DO NOT attempt to hide a reference from the examiner by failing to mention it. If the Patent Office finds out that you did this, you will wind up with an invalid patent and will be in various other forms of hot water as well. The law states that you owe a duty of candor to the Patent Office.

Toward the end of the background information section in Fig. 5-3, there are two paragraphs that state the "objects" of the invention. These paragraphs state the purpose or goal of the invention. *Rule:* If you can, you should write these so that the "objects" are attained by your invention but not by the prior art inventions. For example, in Fig. 5-3, one object of the invention is to provide "greater" density, not "high" density or merely "great" density. Someone could say that previous inventions allowed for "high" or "great." However, the object of "greater" density was not achieved in the past, because "greater" can be taken to mean greater than would be the case without this invention.

Fig. 5-4 contains another sample Background of the Invention section. In this case, part 2 of the background is entitled "Description of the Prior Art." This is the standard format. Part 2 does not contain the standard boilerplate clauses we saw in Fig. 5-3. This is a matter of preference.

Starting at column 1, line 24 of Fig. 5-4, prior patents are discussed. Notice that each paragraph begins with a description of a prior patent and ends with a sentence regarding some disadvantage inherent in that patent. This is a very rigid pattern of writing . . . but a good one. The examiner can merely glance at the first few words in each paragraph in order to determine what prior art is being discussed.

Figure 5-4 does not contain any statement of "objects" for the invention. Although this also is a matter of preference, it is common for patent application writers to end the background section with a few sentences describing "objects." Sometimes the "objects" are included not in the background section but in the following section—the Summary of the Invention. This, again, is a matter of individual style.

In those cases where you did not get around to doing a patent search, you are limited to describing, in a general way, the problems in the prior art that are dealt with by your invention. It is preferable to discuss the prior art specifically, with exact patent numbers and citations. However, if you don't have the money or time to get a search done, just skip it. Discuss in general terms what you know about the prior art and describe the problem that your invention will overcome. See Patent No. 4,462,485 for an example of this technique.

Additional examples for the Background of the Invention section are contained in other patents mentioned in this book. You should read all of these in order to pick up the general style of writing used in patent applications. Not all of these samples conform exactly to either the simplified or

recommended format. The recommended format is of recent origin. Accordingly, the older a patent is, the less likely that it will be in the currently recommended form.

Summary of the Invention

The Summary of the Invention explains precisely what invention is being claimed. It should capture the essence of the invention in a single paragraph such that it is made to seem different from previous inventions. This is the Patent Office rule:

37 CFR 1.73 Summary of the Invention. A brief summary of the invention indicating its nature and substance, which may include a statement of the object of the invention, should precede the detailed description. Such summary should, when set forth, be commensurate with the invention as claimed and any object cited should be that of the invention as claimed.

This regulation suggests that you should not write a Summary of the Invention in vague or general terms, such as:

"This invention is a fuel cell which is more economical to run than previous fuel cells."

This violates rule 73 because the stated object is too broad. The object must be commensurate, i.e. have the same scope as the invention. The invention can't be just a fuel cell, because fuel cells have been around for a long time, and it is not an invention to merely do something that is old. The invention must be a specific type of fuel cell that is constructed in a new way. This is the true nature and substance of the invention. Here is better language:

"This invention is a solid electrolyte fuel cell that operates with a wider variety of hydrocarbon fuels than previous cells. This is achieved in a cell comprising a solid electrolyte of porous ceramic impregnated with an alloy comprising zinc and platinum."

Although it is common to state the "objects" of the invention at the end of the Background section, rule 73 suggests that the most appropriate place might be at the very beginning of the Summary.

Figures 5-2, 5-4, 5-5, and 5-6 are patents containing Summary of the Invention sections. Because the Summary contained in an issued patent is exactly as it appeared in the patent application, these can be used as guides for writing summaries for your own applications.

Study these carefully to get a feel for the level of detail needed. Keep this point in mind: the summary of the invention is often the first and last thing people read. They look to it to learn quickly (1) what the invention is supposed to accomplish (i.e. the object) and (2) how it works. The summary explains general theories, goals, and principles of operation. Fine details are saved for the Detailed Description section of the application.

ADJUSTABLE FOLDING TABLE AND HANGER

BACKGROUND OF THE INVENTION

1. Field of the Invention

The present invention relates to tables having foldable table tops which are foldable into a compact unit.

2. Description of the Prior Art

Many times it is desirable to have a table for use during leisure outdoor activities. Such a table should be easy to transport and require little effort to set up. Normally, such a table would be a foldable table, foldable to an easy carrying form. Foldable tables are well known in the art. However, the tables disclosed in the prior art do not provide a versatile table that is amenable to a range of various outdoor activities, such as fishing at a river or lakeshore or simply sitting and enjoying the outdoors in one's backyard. In either of the above examples, a table is useful for placement of various articles such as fishing tackle or food and refreshments. Some examples of the prior art foldable tables and other structures for supporting articles are set forth in the patents briefly described below.

The Sutton U.S. Pat. No. 2,217,031 shows a table that is collapsible by first folding the feet which support the center post vertically and then folding the center post about a pivot with respect to the table top. Although the table of the Sutton Patent is foldable, the manner in which the table is foldable makes the table awkward to carry.

The Zielfeldt U.S. Pat. No. 2,750,243 also shows a table which has four legs hinged to the top of the table such that the table is folded. The table further has a hand hold on one edge for carrying the table in a folded position. However, the table of the Zielfeldt Patent is awkward in many outdoor situations due to the four legs.

The Futrell U.S. Pat. No. 3,638,585 shows a combination lectern and table. The table has two halves, one half being foldable along the supporting post and the other half is removable. The lectern and table of the Futrell Patent would be awkward to carry due to the removable half of the table top.

The Sutton U.S. Pat. No. 103,387 discloses a table with four equally spaced radially extending support arms. The support arms support four individual table tops, each top being pivotal downwardly independent of the other table tops. Although the table tops are foldable with respect to the main support, the table shown in the Sutton Patent would be awkward to carry since the radially extending support arms still extend horizontally from the center post.

The Coblentz U.S. Pat. No. 861,810 shows a wash stand having a base supported by feet at a central post extending upwardly from the base. A table top is pivotally attached at the top of the support post and pivots outwardly when a support arm is detached from the underside of the table top. The wash stand, however, would be awkward to use in most outdoor types of activities where a table would be useful.

The Brandenburg U.S. Pat. No. 2,137,799 and the Jones U.S. Pat. No. 3,177,825 both illustrate foldable chairs. Although the chairs fold into a generally compact structure for carrying, they have little value as a table for supporting articles thereon.

The Lorton U.S. Pat. No. 1,666,293 illustrates a campfire outfit having a central post with a lower portion for insertion into the ground. Various articles are then attached to the central post, such as a rack, on which other articles may be placed or food cooked.

SUMMARY OF THE INVENTION

The present invention includes a table having a substantially vertical support member with a lower ground engaging portion. First and second leaves form a table top and are pivotally attached to each other. A support linkage supports the first and second leaves about the vertical support and is pivotally attached to the underside of the leaves at an upper end and is pivotally and slidably connected to the vertical support member. The support linkage and the support member are disposed within the same plane when the leaves are pivoted to a down position for folding the table into a compact and easily transportable unit. Preferably, a hanger is attached to an upper end of the vertical support member and is adjustable to a desired height above the table top.

Fig. 5-4. Patent Extract: Adjustable Folding Table and Hanger.

Brief Description of the Drawings

There should be one sentence for each figure. This sentence should say (1) what the figure shows and (2) how it shows it i.e. whether it is a cut-away, exploded, perspective view, etc. Do not refer to numbered parts of the figures or explain how the invention works.

Detailed Description of the Invention

The detailed description provides details regarding how to make and use the invention. The patent statutes require that the description teach the "best mode" for practicing, making, and using the invention. If substantially different versions of the invention are also to be protected, details regarding these must also be disclosed.

The Summary of the Invention covers the goals or objects of the invention and briefly describes how it works. The Detailed Description section refers to the drawings, explains what each component in the drawing is, and explains how all of the components work together to provide the end result. It provides details regarding materials used to make the invention and, if it is not obvious, instructions for building the invention.

In general, the more information in the Detailed Description, the better. If the invention is an electronic system, say a device that answers the telephone and provides a message using synthetic speech, it is best to give a schematic showing every component and the value of every component, together with listings for any required microcontroller software, and complete instructions for installing and using the device.

The penalty for failing to write a complete detailed description is harsh. If, for example, you do not provide a complete schematic, the examiner might say that the description does not teach how to make and use the invention. If your application is rejected for this reason, it will be very difficult to repair the damage, because additional details cannot be added to your description after the application has been filed. Additional detail is considered to be "new matter," which is forbidden in amendments to the patent application. To get more detail in front of the patent examiner, you might be forced to file a second application and accept a later filing date for the new matter.

Although it is ideal to present a complete component level schematic of your invention, it is sometimes impossible to do this. Perhaps you have not had time to build an actual device or have not yet written the required microcontroller software. Under these circumstances, a patent application can still be filed if you are willing to accept a degree of risk. Glossing over very conventional details does not generate a great deal of risk. For the the telephone answering machine example, items such as power supplies or audio amplifiers could be shown as simple functional blocks in drawings and in the written description. It is even possible to indicate a single chip computer as a block, especially if it is accompanied by a flow chart of the program the chip is to execute with connections for the I/O pins indicated. If a technician could use

your description and drawings to make the invention without undue experimentation, the description requirement is satisfied. If all the subsystems in the invention are conventional and it consist of a new way of putting these subsystems together to perform a new function, then you can, in theory, omit a component-level schematic. However, if any of the subsystems require unique circuitry or other construction to work properly, provide at least a component-level schematic of this portion.

A DETAILED DESCRIPTION HORROR STORY

One reason for putting as much information as possible into a detailed description is that there are sometimes things about inventions that even the inventor does not understand.

One inventor of a chemical process described all of the steps in his process for making a new kind of dye, but he neglected to mention that he used an iron pot to hold the chemicals during one of the heating steps. The inventor wasn't aware of what difference the type of pot could make. So why mention iron in the detailed description? Unfortunately for this inventor, it was later shown that the iron acted as a catalyst. His process failed unless iron was present during that particular step. Because he did not mention iron in his detailed description, it did not teach "how to make and use" the invention, and a court held the patent to be invalid. Moral of the story: do not lightly assume that any particular detail is irrelevant. When in doubt, put the details in!

Figures 5-2, 5-4, 5-5, and 5-6 also contain well-written Detailed Description sections. Pay particular attention to the way these descriptions constantly refer to the accompanying drawings. Notice that components shown in drawings are labeled with numbers and that these numbers are mentioned in the detailed description. This is required by Patent Office rules.

A SOFTWARE PATENT

Figure 5-5 is something very important: a patent for software. Notice that in addition to the detailed description, an appendix with a complete program listing is attached. By including a program listing with his patent application, the inventor made sure that he met the requirement that his detailed description teaches how to make and use the invention. Figure 5-2 also contains an excellent detailed description for a simple mechanical device, a soccer practice machine.

According to the patent statute, a detailed description of the invention must provide details sufficient to enable any person skilled in the pertinent art or science to make and use the invention without undue experimentation. The knowledge of those skilled in the art will vary widely depending on the technology you are dealing with. To be on the safe side, assume that the person reading your detailed description is a technician or craftsman who can follow instructions and make something—not an engineer or physicist who can figure it out for himself.

The way you write the detailed description will vary depending upon whether you are claiming a process, machine,

United States Patent [19]

Goldwasser et al.

[11] **Patent Number:** **4,559,598**

[45] **Date of Patent:** **Dec. 17, 1985**

[54] **METHOD OF CREATING TEXT USING A COMPUTER**

[76] Inventors: **Eric Goldwasser; Dorothy Goldwasser,** both of 993 Barberry Rd., Yorktown Heights, N.Y. 10598

[21] Appl. No.: **468,493**

[22] Filed: **Feb. 22, 1983**

[51] Int. Cl.⁴ G06F 15/21
[52] U.S. Cl. 364/419; 364/900
[58] Field of Search .... 364/419, 200, 900; 340/707, 708, 365 P; 434/167, 169, 308; 400/98, 95

[56] **References Cited**

U.S. PATENT DOCUMENTS

4,190,833	2/1980	Bejting et al.	340/707
4,193,119	3/1980	Arase et al.	364/900
4,195,338	3/1980	Freeman	364/200
4,198,623	4/1980	Misek et al.	340/365 P
4,247,906	1/1981	Corwin	364/900
4,386,232	5/1983	Slater	178/18
4,406,626	9/1983	Anderson et al.	364/419
4,438,505	3/1984	Yanagiuchi et al.	364/419
4,456,973	6/1984	Carlgren et al.	364/900
4,464,070	8/1984	Hanft et al.	400/98
4,481,603	11/1984	McCaskill et al.	364/900

OTHER PUBLICATIONS

Cossalter et al., "A Microcomputer Based Communication System for the Non-Verbal Severely Handicapped" Eurocon '77 Proceedings on Communications, Venice, Italy, May 1977, pp. 196–202.

Primary Examiner—Michael R. Fleming
Attorney, Agent, or Firm—Karl F. Milde, Jr.

[57] **ABSTRACT**

A method of creating text using a computer having a display screen and a pointing mechanism for identifying locations on the display screen. The method involves displaying a list of commonly used words on the screen so that the user may select the words to be used in the text by successively pointing to them. The computer responds to the pointing mechanism and displays on the screen a line of text comprising the identified words in their successive order of selection.

11 Claims, 2 Drawing Figures

```
    this is the eleventh line of text.
     Now is the time for all good men to
    come to the aid of their country.


     a   each if   on    so    up   now
    all       in   of    some       time
    and for   is   one         very good
    are from  it   or    that        men
    as             out   the   was  come
    at  go   just        their we   aid
                  put     them  were country
    be  had  know        then  what
    but have             there when
    by  he   like        these will
        her       right  they  with
    can his  make        this  word
        how  my   said   to
    do             see   too   you
    did I    not   she   two   your

    Q W E R T Y U I O P  1 2 3 4 5 6 7 8 9 0

    A S D F G H J K L   end  Cap Par s ed ing

    ! Z X C V B N M , . ? ' " - - edit erase
```

Fig. 5-5. Patent: Method for Creating Text Using a Computer.

```
this is the eleventh line of text.
 Now is the time for all good men to
come to the aid of their country.


a    each if  on     so    up   now
all       in  of     some       time
and  for  is  one          very good
are  from it  or     that        men
as            out    the   was  come
at   go   just       their we   aid
              put    them  were country
be   had  know       then  what
but  have      quiet there when
by   he   like       these will
     her       right they  with
can  his  make       this  word
     how  my   said  to
do             see   too   you
did  I    not  she   two   your


Q W E R T Y U I O P  1 2 3 4 5 6 7 8 9 0

A S D F G H J K L  end  Cap Par s ed ing

! Z X C V B N M , . ? ' " - - edit erase
```

FIG. 1

```
this is the second line. As you might
have guessed this is the third line. The
fourth line is, you got it, this very
one. By this time I don't have to tell
you what line this is because you
probably can figure it out for yourself.
But I want to direct your attention to
the following line which, if I counted
correctly, is the tenth line. And then
this is the eleventh line of text.
 Now is the time for all good men to
come to the aid of their country.

ADD_WORDS   ALIGN   PRINT  SAVE   RETURN

INSERT_LETTER  INSERT_WORD    TOP    UP

DELETE_LETTER  DELETE_WORD  BOTTOM  DOWN

Q W E R T Y U I O P  1 2 3 4 5 6 7 8 9 0

A S D F G H J K L  end Cap Par (@#$%&*)

! Z X C V B N M , . ? ' " - - :;-/ erase
```

FIG. 2

METHOD OF CREATING TEXT USING A COMPUTER

BACKGROUND OF THE INVENTION

This invention is a new way of creating text using a pointing mechanism such as a light pen or touch sensitive display screen to point to words that a program causes to be displayed on the screen, thereby adding them to the text.

The standard method of creating text is by typing on a keyboard. This method requires considerable skill to be able to create text at reasonably high rates. It typically takes adults about 100 hours of training to attain a typing speed of about 40 words per minute. Young children are not taught to type because they do not have the mental and/or physical dexterity required.

As personal computers become less expensive in the next few years, we can expect that almost anyone who needs to create text will have access to a computer so that he can use the word processing capabilities of the computer to create the text. Further, since word processing offers great advantages over both handwriting and standard typing on paper via a standard typewriter, we can expect that using a computer will be the method of choice for creating text. Hence a method of creating text at a reasonably high rate using a computer, that requires little skill or training will be very advantageous.

SUMMARY OF THE INVENTION

An object of this invention is to provide a method of creating text using a personal computer, which method is easy to learn but holds the possibility of permitting both young children and adults to create text at higher rates than present methods permit.

Another object of this invention is to provide a method of creating text which allows a person to use a word even if he is not sure of the correct spelling of that word.

Still another object of this invention is to provide a method of editing using a natural pointing method rather than typed commands or cursor movements.

These objects, as well as other objects which will become apparent from the discussion that follows, are achieved, according to the present invention, by suitably programming a computer, having both a display screen and a pointing mechanism for identifying locations on the display screen, to enable the computer to carry out a process with the following steps:

(a) displaying on the screen a list of language words for selection by the computer user;

(b) identifying successive ones of the words in response to the pointing mechanism as these words are selected by the user by pointing to their respective locations on the screen; and

(c) displaying on the screen a line of text comprising the identified words in their successive order of selection.

Thus, when a person uses the programmed computer to create text on the display screen, he (or she) points to successive ones of the listed words, which are preferably arranged alphabetically on the screen. The computer then notes which words were pointed to and performs the appropriate processing to display a line or lines of text containing these words in proper order. Advantageously, the line(s) of text can be displayed at either the top or the bottom of the screen, so as to leave

room on the main part of the screen for the original "library list" of words.

According to a particular feature of the present invention a list of alphanumeric characters is also displayed, so that words which are not contained in the "library list" may be spelled out, character by character, by pointing to their successive characters, in turn, in the manner described below.

Typically the first screen that is displayed (i.e. the first set of words and characters) by the computer will contain the most common words so that the person can add one of these words to the text by pointing to it, with one pointing action. If the person wants to add a word that is not on the first screen, he can point to one or more words or characters that will cause the computer to display a new screen on which the word that he wants to add is present. Typically he would point to the first letter of the word he wants. The computer will then display a second screen of words all of which start with that letter and which are in some sense the most common words that start with that letter. Hence this second screen is likely to contain the word that he has in mind. If it does, he can add the word to the text by pointing to it. He will then have added that word with a total of two pointing actions. If this second screen does not contain the word he wants, he would point (on this second screen) to the second letter of the word he wants and the computer will then display a third screen of the most common words that start with the first letter followed by the second letter. He can continue spelling out the word by pointing to succeeding letters until a screen is displayed which contains the word he wants or he has spelled out the entire word. If the word appears he can point to it and it will be added to the text and the first screen will be displayed so that he can choose his next word from among the most common words. If the word is completely spelled out he can point to an end of word indicator and the word will be added to the text and to one of the screens so that the next time he wants to use that word he will not have to spell it out again.

To edit the text he can point to an edit indicator and the computer will display a portion of the text along with edit command words. He can then perform the editing functions (such as "delete", "insert", etc.) by pointing to the edit command words and the text words that the edit commands should operate on.

The pointing mechanism can be a light pen, or a touch sensitive display that registers the position on the screen that a fingertip or other pointing object touches, or some other way of selecting a point on the screen. The computer can also be constructed to register several points that are pointed to almost simultaneously by several fingers or other pointing objects. Hence a person could point to the first two letters of a word almost simultaneously with two fingers, or he could point to a letter and a position on the screen at which he expects a word to appear as a result of pointing to the letter. This method according to the invention makes it possible for people to create text at very high rates.

The foregoing and other objects, features and advantages of the invention will be apparent from the following more particular description of the preferred embodiments of the invention, as illustrated in the accompanying drawing.

BRIEF DESCRIPTION OF THE DRAWING

FIG. 1 of the drawing illustrates a preferred format for the display screen of a computer which is programmed to operate according to the method of the present invention.

FIG. 2 illustrates an edit screen format for use with the method of the present invention.

DESCRIPTION OF THE PREFERRED EMBODIMENT

The preferred embodiment of the present invention will now be described in connection with FIGS. 1 and 2 of the drawing. This embodiment utilizes the following computer equipment, which is available commercially from IBM Corp., Boca Raton, Fla.:

IBM Personal Computer with at least 64K of memory;

At least one diskette drive;

A graphics CRT display;

A light pen.

The preferred embodiment also utilizes a computer program, written in BASIC language and designed to be supported by the IBM Disk Operating System (DOS). A complete listing of the instructions of this program is set forth in the attached Appendix. This program is entitled "Pointwriter" ™.

The Pointwriter ™ text processing program operates to carry out 25 different functions. These are:

1. Add word to text and most used words columns.
2. Add letter to partially spelled word and get new screen of words.
3. Add fully spelled word to text and most used words columns.
4. Add suffixes "s", "ed", and "ing".
5. Add other suffixes.
6. Add special characters (including numbers and punctuation).
7. View text.
8. Replace character(s).
9. Delete character(s).
10. Insert character(s).
11. Delete word(s).
12. Undo last action.
13. Insert word(s).
14. Capitalize.
15. Paragraph.
16. Save text and most used words columns.
17. Print text.
18. Align text to eliminate short lines.
19. Stop editing and return to main screen.
20. Stop edit function.
21. Use words that you have previously used.
22. Start a new document or add to an old document.
23. Use old most used words columns.
24. Modify screens (for teacher).
25. Tutorials.

1. Add word to text and most used words columns.

The Pointwriter program generates a main image on the display screen in the format illustrated in FIG. 1. This image or "screen" is divided into three areas:

The first three rows contain the last three lines of text;

The bottom three rows (actually the bottom five because two are blank) contain the alphabet, special characters and command words; and

The middle area consists of a left part of fifteen rows by twenty-nine columns and a right part of fifteen rows by ten columns. The left part contains seventy of the most commonly used English words. The right part contains the fifteen text words that

4

were used most often in the text, but are not among the seventy.

To add a word that is on this screen (except for the command words) to the text you just point to the word. All the characters to the left and right, including the character pointed to, up to the first space will be added to the text. Then a space character will be added to the text so that the next word added will be separated from this word by a space. If the word is not one of the seventy, it may also be added to the right hand column of fifteen most often used words. More precisely, its new frequency (after incrementing its frequency by one) will be compared to the frequencies of the words in that column and if its frequency outranks one of the others' it will replace it. If there are less than fifteen words in the column it will be added to the column. There are also twenty-five other columns of most used words, one for each letter except "x". These columns appear, as we shall see when we discuss the next function-"Add letter to partially spelled word . . .", when the middle area of the main screen is replaced by a list of words all starting with the letter. The word will also be added to the column of fifteen most used words which begin with its first letter.

2. Add letter to partially spelled word and get new screen of words.

If a word that you want to add to the text is not on the main screen of seventy words or in the column of fifteen most used text words you should point to its first letter in the screen displayed keyboard on the left side of the bottom three rows. The words in the middle area of the screen will immediately be replaced with words that begin with that letter. On the left (fifteen row by twenty nine character) part will be between ten and fifty very common words that begin with that letter. On the right (fifteen row by ten character) part will be up to fifteen words that have been used most frequently in the text that begin with that letter. If the word you want is still not on the screen you should point to its second letter. Again the words in the middle area will be replaced with words that begin with these two letters (the first followed by the second). If you point to a third or fourth letter one of two things will occur. If there are many words that begin with that sequence of three or four letters then a new set of words all starting with that sequence will appear. But if there are not many such words then all the words that do not begin with that sequence will be blanked out so that it will be easier to see if your word is on the screen. As you continue to spell out a word the words that don't begin with the letters that you have pointed to will continue to be blanked out so that if you fully spell the word only that word (and perhaps additional words that differ from it by an added suffix) will remain. As you spell out a word the letters you point to are added to the text so that you can see how much of it you have spelled. At any point that you notice the word and point to it, it will be added to the text replacing the partially or fully spelled word. Then the first screen of seventy words will reappear so you can choose your next word.

3. Add fully spelled word to text and most used word columns.

If you have fully spelled a word because it was never displayed, you must point to "end" on the next to last row to tell the program that it has no more letters. It will then be added to the text and possibly to the columns of fifteen most used words. Sometimes it will go into the column of most used words on the first screen.

5

Usually it will go into the column of fifteen most used words that begin with its first letter so that when you want to use it again it will appear on the screen after you point to its first letter.

4. Add "s", "ed", and "ing" to the last word of text.

By pointing to "s", "ed", or "ing" on the right side of the next to last row, these suffixes may be added to the last word displayed.

5. Add other suffixes to the last word of text.

By first pointing to the left arrow on the bottom row to position the cursor immediately after the last text word, and then spelling the suffix and pointing to "end" other suffixes may be added. The full word including the suffix will be ranked by frequency of use for possible inclusion in the most used words columns.

6. Add special characters (including numbers and punctuation).

First position the cursor if necessary by pointing (perhaps repeatedly) to the arrows on the bottom row, then if the character is on the main screen point to it. If it is not on the main screen you should first point to "edit" on the bottom row. The edit screen will appear. Point to the character and you will immediately be returned to the main screen.

An example of the edit screen is shown in FIG. 2. Notice that the last twelve lines of text are displayed along with three additional rows of commands and that special characters have replaced the suffixes and the "edit" command on the bottom two rows.

7. View the text.

Only the last three lines of text are displayed on the main screen. To view the rest of the text point to "edit" to get the edit screen. You will initially see the last twelve lines of text. Point to TOP, BOTTOM, UP, and DOWN to scroll the text. TOP will display the first twelve lines of text and BOTTOM the last twelve lines. UP and DOWN cause a scroll of one line. Pointing to the number n immediately after pointing to UP or DOWN will cause an additional scroll of n-1 lines.

8. Replace a character.

You may replace a character by getting to the edit screen and first pointing to the character you want to replace. The cursor will move to that character. Then point to the character in the last three rows that you want to replace it with. Alphabetic characters will go in as lower case. To make it upper case capitalize it. After the replacement the cursor will move one character to the right so you can continue replacing characters easily.

9. Delete a character.

You may delete a character by getting to the edit screen and first pointing to the character and then pointing to DELETE LETTER. The characters to the right will be moved left one character so you can delete additional characters by repeatedly pointing to DELETE LETTER.

10. Insert a character.

You may insert a character by getting to the edit screen and pointing to the character before which the insertion should take place and then pointing to INSERT LETTER. Then point to the character you want to insert. It will go into the cursor position and the character that was at that position and the characters to the right will move one character to the right as will the cursor so you can easily continue inserting characters.

11. Delete words.

6

You may delete words by getting to the edit screen and pointing to DELETE WORD. Any words that you point to after that will be deleted.

12. Undo your last action.

You may undo your last action by pointing to "erase" on the bottom row.

13. Insert words.

You may insert words by getting to the edit screen and pointing to INSERT WORD. You will be asked to point to the word before which you want to insert. After you point to that word the main screen will reappear with the top three rows containing the text before that word on row one and the text after that word on row three and only the cursor on row two. You can insert as many words as you want just as you would add words to the end of the text. When you are finished inserting, point to the end. The text is aligned and then the last three lines of text will appear and you can continue.

14. Capitalize a character, by pointing to "Cap" and then the character. On the main screen you must point to "Cap" before each character, but on the edit screen you can point to "Cap" once and then a succession of characters.

15. Start a new paragraph, by pointing to "Par". On the main screen this will put the cursor at the beginning of the next text line indented one character. On the edit screen you will be asked which word should start a new paragraph and when you point to a word a new paragraph will be created and the text will be re-aligned accordingly.

16. Save your document on disk, by pointing to SAVE on the edit screen. You will be asked to spell a title. You should point to "end" after spelling a title. The columns of most used words will also be saved under this title.

17. Print your document on the printer, by pointing to PRINT on the edit screen.

18. Eliminate short lines caused by deletions, by pointing to ALIGN on the edit screen.

19. Stop editing and return to the main screen, by pointing to RETURN on the edit screen.

20. Stop edit functions, by pointing to "end" on the edit screen. In some cases you can point to another edit command to automatically stop the previous edit function, but in the case of DELETE WORD you must point to "end" before attempting to position the cursor for another edit function since pointing to a text word while in DELETE WORD mode will delete the word!

21. Use words that you have previously used, by pointing to them in the text or the columns of most used words. If the word you want to use is in the last three lines of text it will be in the main screen text area and you can add it to the text by pointing to it. If it is not in the last three lines you can go to the edit screen and point to ADD WORDS. Then when you point to text words they will be added to the text instead of the cursor being positioned at the word.

22. Start a new document or add to an old one, by pointing to NEW DOC or to one of the titles that is displayed when you start the Pointwriter program. The initial Pointwriter program screen contains instructions and the command START. When you point to START the titles of previously saved documents will be displayed.

23. Use old most used words columns.

After you point to a title you will be asked if you want to use the most used words columns associated with that document or start out fresh. If you start a new document you will be asked if you want to use the most used words columns associated with one of your stored documents.

24. Modify word screens.

The teacher can modify any of the word screens (of fifteen rows by twenty nine characters) by adding and-/or deleting words. The new screen can then be saved in place of the supplied screen.

25. Tutorials

There are two tutorials. One presents a story and asks a child to recreate it. As the child successfully recreates the words they are highlighted so the child can see

which word to work on next. The other presents a story and recreates it itself showing the child the proper pointing actions by highlighting the words, letters, and command words in the proper sequence.

There has thus been shown and described a novel method of creating text which fulfills all the objects and advantages sought therefor. Many changes, modifications, variations and other uses and applications of the subject invention will, however, become apparent to those skilled in the art after considering this specification and the accompanying drawing which discloses preferred embodiments thereof. All such changes, modifications, variations and other uses and applications which do not depart from the spirit and scope of the invention are deemed to be covered by the invention which is limited only by the claims which follow.

APPENDIX

```
1 'PointWriter (C) Eric Goldwasser, January, 1983
5 'ON ERROR GOTO 10000
10 'initialize
20 KEY OFF:WIDTH 40:CLS:LOCATE 1,1
21 PRINT " Hello, my name is PointWriter. You can"
22 PRINT "use me to write stories by pointing to"
23 PRINT "words with the light pen. If you want"
24 PRINT "to use a word that is not on the screen"
25 PRINT "just point to the first letter and a"
26 PRINT "new screen will appear. The word might"
27 PRINT "be on the new screen. If the word is"
28 PRINT "not on the new screen then point to the"
29 PRINT "second letter and a new screen will"
30 PRINT "appear. You might have to spell the"
31 PRINT "word by pointing to all its letters."
32 PRINT "Then you should point to end' to tell"
33 PRINT "me that there are no more letters in"
34 PRINT "the word."
35 PRINT " I am loading my memory with words."
36 PRINT "Please wait for me to finish loading my"
37 PRINT "memory. I will beep when I'm finished."
40 DEF SEG=&HBBA0:BLOAD"DISPLAY",0:DISPLAY%=0
42 DEF SEG=&H3C00:BLOAD"alpmain",0
44 SCRN$="x":R%=21:C%=1:O%=0:L%=400:SH%=&H3C:SL%=0:DEF SEG=&HBBA0
46 CALL DISPLAY%(SH%,SL%,O%,L%,R%,C%)
50 DEF SEG=&HBBB0:BLOAD"fprint",0:FPRINT%=&H100
51 DEF SEG=&HBBC0:BLOAD"rnk12",0:RNK12=&H200
52 DEF SEG=&H3F00
53 FOR I=0 TO &H557:POKE I,32:NEXT
54 FOR I=&H90 TO &H9B:POKE I,122:NEXT
55 FOR I=&H558 TO &H5C9:POKE I,0:NEXT
56 POKE &H564,122
57 POKE &H5C9,12
59 DEF SEG=&HBBE0:BLOAD"prt15",0:PRT12=&H400
60 DEF SEG=&H1800:BLOAD"az",0
61 DEF SEG=&H2000:BLOAD"abwr1",0
62 DEF SEG=&H2A00:BLOAD"abwr2",0
63 DEF SEG=&H3800:BLOAD"canthe",0
64 DEF SEG=&H3B00:BLOAD"coxx",0
67 DEF SEG=&H3D00:BLOAD"edscrn",0
68 DEF SEG=&H3C00:BLOAD"alpmain",0
70 'entry
71 UCALPH$="ABCDEFGHIJKLMNOPQRSTUVWXYZ"
72 TWOLET1$="ab ac ad af ag ai al am an ap ar as at au av aw ba be bi bl bo br b
u ca ce ch ci cl co cr cu cy da de di dr du ea el em en es ev ex fa fe fi fl fo
fr fu ga ge gi gl go gr gu ha he hi ho hu hy id il im in la le li lo lu ma me mi
mo mu my"
73 TWOLET2$="na ne ni no nu ob oc of on op or ou pa pi pl po pr pu ra re ri ro r
u sa sc se sh si sk sl sm sn so sp sq st su sw ta te th ti to tr tu tw ty wa we
wh wi wo wr"
74 THRLET$="can car com cox cou ins int inv par pre pro rea rel res str the"
75 DIM TXT$(100):DIM S15$(20):DIM OTXT$(100)
```

```
12260 PRINT #2,Z15:FOR I=1 TO Z15:PRINT #2,Z15$(I):NEXT
12900 LOCATE 13,25:PRINT "It's been fun"
12999 END
18000 CLOSE #1
18002 OPEN "col15.txt" FOR INPUT AS 2
18010 INPUT#2,A15:FOR I=1 TO A15:INPUT#2,A15$(I):NEXT
18020 INPUT#2,B15:FOR I=1 TO B15:INPUT#2,B15$(I):NEXT
18030 INPUT#2,C15:FOR I=1 TO C15:INPUT#2,C15$(I):NEXT
18040 INPUT#2,D15:FOR I=1 TO D15:INPUT#2,D15$(I):NEXT
18050 INPUT#2,E15:FOR I=1 TO E15:INPUT#2,E15$(I):NEXT
18060 INPUT#2,F15:FOR I=1 TO F15:INPUT#2,F15$(I):NEXT
18070 INPUT#2,G15:FOR I=1 TO G15:INPUT#2,G15$(I):NEXT
18080 INPUT#2,H15:FOR I=1 TO H15:INPUT#2,H15$(I):NEXT
18090 INPUT#2,I15:FOR I=1 TO I15:INPUT#2,I15$(I):NEXT
18100 INPUT#2,J15:FOR I=1 TO J15:INPUT#2,J15$(I):NEXT
18110 INPUT#2,K15:FOR I=1 TO K15:INPUT#2,K15$(I):NEXT
18120 INPUT#2,L15:FOR I=1 TO L15:INPUT#2,L15$(I):NEXT
18130 INPUT#2,M15:FOR I=1 TO M15:INPUT#2,M15$(I):NEXT
18140 INPUT#2,N15:FOR I=1 TO N15:INPUT#2,N15$(I):NEXT
18150 INPUT#2,O15:FOR I=1 TO O15:INPUT#2,O15$(I):NEXT
18160 INPUT#2,P15:FOR I=1 TO P15:INPUT#2,P15$(I):NEXT
18170 INPUT#2,Q15:FOR I=1 TO Q15:INPUT#2,Q15$(I):NEXT
18180 INPUT#2,R15:FOR I=1 TO R15:INPUT#2,R15$(I):NEXT
18190 INPUT#2,S15:FOR I=1 TO S15:INPUT#2,S15$(I):NEXT
18200 INPUT#2,T15:FOR I=1 TO T15:INPUT#2,T15$(I):NEXT
18210 INPUT#2,U15:FOR I=1 TO U15:INPUT#2,U15$(I):NEXT
18220 INPUT#2,V15:FOR I=1 TO V15:INPUT#2,V15$(I):NEXT
18230 INPUT#2,W15:FOR I=1 TO W15:INPUT#2,W15$(I):NEXT
18250 INPUT#2,Y15:FOR I=1 TO Y15:INPUT#2,Y15$(I):NEXT
18260 INPUT#2,Z15:FOR I=1 TO Z15:INPUT#2,Z15$(I):NEXT
18800 CLOSE #2
18900 GOTO 90
```

What is claimed is:

1. A method of creating text using a computer having a memory, at least one display screen and means for selecting positions on said screen, said method comprising the steps of:

(a) storing in said memory a dictionary of frequently used linguistic expressions, at least some of said linguistic expressions having a plurality of alphanumeric characters;

(b) displaying on a first section of said screen a plurality of said linguistic expressions arranged in a predetermined order for selection by a user;

(c) displaying on a second section of said screen at least one line of text, as said text is created by a user;

(d) identifying the position of a linguistic expression on said first section of said screen, in response to selection of that position by a user with said position selecting means; and

(e) displaying the linguistic expression, whose position was identified in step (d), in said second section of said screen, concatenated to the end of said line of text, thereby adding a linguistic expression to said line of text.

2. The method defined in claim 1, wherein some of said linguistic expressions each comprise a single alphanumeric character.

3. The method defined in claim 2, wherein said single alphanumeric characters are displayed on said first section of said screen arranged in the order they appear on a "QWERTY" keyboard.

4. The method defined in claim 1, wherein said predetermined order includes alphabetic order.

5. The method defined in claim 1, wherein step (b) includes the step of displaying a list of words of a given language, which are the words most likely to be selected by the user.

6. The method defined in claim 1, further comprising the additional step of identifying, in sequence, at least one alphanumeric character selected by a user, and wherein step (b) includes the step of displaying on said first section of said screen a plurality of said linguistic expressions which begin with said at least one alphanumeric character identified in said additional step in the character sequence selected by the user.

7. The method defined in claim 6, wherein said at least one alphanumeric character is selected by a user by typing on a keyboard, and wherein said additional step includes the step of receiving typed commands from said keyboard.

8. The method defined in claim 6, wherein some of said linguistic expressions each comprise an alphanumeric character and wherein said at least one alphanumeric character is selected by a user by identifying its position on said first section of said screen with said position selecting means.

9. The method defined in claim 1, wherein said means for selecting positions on said screen comprises a pointing mechanism.

10. The method defined in claim 9, wherein said pointing mechanism is a light pen.

11. The method defined in claim 9, wherein said pointing mechanism is a touch-sensitive display.

* * * * *

manufacture, composition of matter, or improvement. In the case of a chemical invention, the detailed description should include a wide range of specific examples, if possible. Because chemistry is a rather uncertain, empirical science, a wide range of actual, tested examples will help support a broader patent. For other types of invention, a single example is often sufficient.

The specification should be written in common English without jargon. Although the patent examiner can understand highly technical descriptions, a patent could someday become contested in court. Here, it will be read by a judge and jury who are not technically oriented. Management personnel at corporations who are thinking about buying rights to a patent will also appreciate clear, non-technical language.

SAVE DISCUSSION OF ALTERNATIVES FOR LAST

Choose one embodiment or sample of the invention and start off the detailed description by describing this thoroughly. Do not get bogged down at the beginning of the detailed description by trying to describe every possible variation or substitution that could be made in the construction of the invention. Save a discussion of alternative and variables for a separate section in the last part of the description. A good sentence for starting the detailed description is this:

"Refer now to Fig. 1, which is an overall drawing of a preferred embodiment of the invention."

This introductory sentence makes it clear to the reader that you are describing just one example, and that this is just a *preferred* way to implement the invention—not the only way.

When writing a detailed description, explain how the parts of the invention work together to produce a useful result. Do not merely list or catalogue parts. If the invention pertains to a process, machine, or electrical circuit, describe step-by-step how the thing operates. Refer to the numbers in the drawings. Be sure to refer to and discuss every drawing. Number the important points in each drawing and describe each part in the text. Use the same number to refer to the same part in all drawings.

Explain the theory behind the invention, if possible. When writing claims for the invention, it is sometimes useful to use the theory of operation as one way of distinguishing the invention from the prior art. However, if unsure of the theory, qualify your description with a statement such as "it is believed that the invention operates as follows . . . ". If it later turns out that your ideas about the theory of the invention are wrong, this language will prevent others from saying that there is misleading or fraudulent language in the patent.

After completely describing one preferred version of the invention, proceed to mention or describe other embodiments or alternate versions.

In the second to last paragraph of the detailed description, summarize the detailed description and state how the invention has fulfilled the objects of the invention that were stated previously.

The last paragraph of the detailed description should be a boilerplate statement as follows:

"The foregoing description of the preferred embodiment of the invention has been presented for the purposes of illustration and description. It is not intended to be exhaustive or to limit the invention to the precise form disclosed. Many modifications and variations are possible in light of the above teaching. It is intended that the scope of the invention be limited not by this detailed description, but rather by the claims appended hereto."

This paragraph says that the invention covers more than just the precise thing described. It is a broad concept, and any device that is within the language of the claims is to be within the coverage of the patent. This prevents others from pointing to specific examples and arguing the patent is limited to these.

THE RULE OF DETAILED DESCRIPTION WRITING

There is one rule concerning the writing of the detailed description that is more important than any other: ALL LANGUAGE USED IN THE CLAIMS SHOULD APPEAR SOMEWHERE IN THE DETAILED DESCRIPTION. Claim writing is discussed in the next section. However, because all claims must be based on the drawings and language of the detailed description, the golden rule has to be mentioned now. After writing the claims, examine the detailed description to be sure that every element of each claim is mentioned and described thoroughly. If this is not done, the patent examiner can reject claims for "lack of basis in the specification."

For example, if claims mention a "diode" but the detailed description contains only the word "rectifier," this rule has been violated. Synonyms do not count. After writing the detailed description and the claims, test your work as follows: On a draft copy of the patent application, use a yellow highlighting marker to mark each phrase in the claims that also appears in the detailed description. If you cannot find a particular phrase in the detailed description, add a sentence to the description that includes the phrase. Then highlight the phrase in the claims section. When you have managed to highlight every phrase in the claims, your detailed description has passed this test.

Patent Claims

Following the detailed description section of the patent application, you must particularly point out and claim exactly what it is that defines the invention. This is the Claims section of the application.

The previously described sections of the application can be written in (more or less) plain English. However, the Claims section must be written in a very formal way, and certain magic words must be used.

The best way to learn how to write claims is to read those written by others. The claims of U.S. Patent 4,605,881, a General Electric patent for a sodium arc lamp, which is reproduced in Fig. 5-6, are discussed in this section.

Claims always start out with the words "What is claimed is" or if you like, "I claim" or "We claim." These words are the beginning of a sentence. The sentence is ended by the remainder of a claim. Introductory words such as these and a single-sentence format are mandatory. Notice that many of the sample claims presented in this book are very long and complex, yet they are not broken up into multiple sentences. Although such complex claims can be difficult to read, they do comply with the single-sentence requirement.

The Claims section is divided into blocks of text, and each block has a number. Periods are used only after the number at the beginning of a block of text, as in "C2.", and only at the very end of a block which is obviously consistent with the single-sentence format. Within a claim, commas or semi-colons must be used to group phrases. Formatting the text into sub-blocks or outline form is helpful, but not required.

After the number of each claim, there is a short introductory phrase that describes the overall subject of the claim. This is called the preamble. For claim 1 of the arc lamp patent, the preamble is "In a high pressure sodium iodide arc lamp having an arc tube for containing an arc discharge, an arc tube fill . . ." For other inventions, you could use preambles such as:

- ✍ "A high speed field-effect transistor . . ."
- ✍ "A mechanical clock . . ."
- ✍ "A frying pan . . ."
- ✍ "A composition for making hair grow . . ."
- ✍ "A method for making potato chips . . ."
- ✍ "An improved method for cooking potato chips in oil wherein the improvement comprises . . ."

In addition to introducing the overall subject of the claim, the preamble is sometimes used to make the claim more specific, i.e. narrower. This helps distinguish the subject matter of your claim from previous inventions. For example, perhaps in the prior art there have been "High speed field-effect transistors . . ." To distinguish the transistors you are claiming, your preamble could be a little more elaborate, such as

"A high speed field-effect transistor which is resistant to radiation damage . . ."

"A high speed field-effect transistor suitable for switching high currents and voltages . . ."

"An improved high speed field-effect transistor with a reduced tendency toward unwanted oscillations . . ."

"A method for making potato chips with a uniform shape and size . . ."

"A method for making frying pans that are easy to wash . . ."

Of course, if adding detail to the preamble allows you to distinguish prior inventions, it also allows other people to distinguish their invention from yours. Specific claims have a better chance of getting approved, but once approved, they cover less territory. A solution to this dilemma is to write numerous claims of differing scope. But his solution has its price: more claims result in a higher filing fee.

The preamble of a claim can also be used to introduce words referenced later in the claim. The arc lamp preamble in Fig. 5-6 begins with "In a high pressure sodium iodide arc lamp having an arc tube . . ." The words "having an arc tube" are included in the preamble so that a later reference to "said arc tube" can be made. If the existence of the arc tube had not been mentioned in the preamble, this reference to an arc tube would introduce confusion. The preamble explains how the arc tube fits into the overall invention and lays a foundation for later reference to the arc tube in the claim. Examiners object to claims that refer to some part of an invention without laying an adequate foundation, i.e. introducing that part in a separate clause. If your claims are objected to for this reason, amend the preamble to include more foundation words.

INDEPENDENT VERSUS DEPENDENT CLAIMS

In the arc lamp claims of Fig. 5-6, notice that the preamble of claim 1 is very different from that of claim 2. Claim 1 is an independent claim—it can stand alone. Claim 2 is a dependent claim that refers to a previous claim and adds some additional limitation. In this case, claim 2 covers only those lamps covered by claim 1 that operate within a range of sodium gas pressure within 10 to 100 torr. Likewise, claim 3 is dependent upon claim 2. Claim 3 covers only those lamps covered by claim 2 that meet an additional limitation regarding iodine gas pressure.

Dependent claims are allowed to cover *only* territory that is included in the independent claim, and furthermore they must cover *less* territory.

For example, it would be invalid to write claim 2 as follows:

2. The lamp of claim 1 wherein chlorine is substituted for iodine.

Because the territory covered by claim 1 includes only lamps with iodine, dependent claim 2 might not be made to cover a lamp that does not contain iodine.

Dependent claims are used because they are a convenient way to add narrower claims without having to repeat the same text. These narrower, dependent claims define a smaller, more easily defended territory to which the applicant may retreat should the larger territory of the broad claims prove indefensible in view of the prior art. The use of dependent claims also results in a lower filing fee.

"COMPRISING" . . . A MAGIC WORD
Notice that the preamble in claim 1 of the arc lamp pat-

United States Patent [19]

Dakin

[11] Patent Number: 4,605,881

[45] Date of Patent: Aug. 12, 1986

[54] **HIGH PRESSURE SODIUM IODIDE ARC LAMP WITH EXCESS IODINE**

[75] Inventor: **James T. Dakin,** Schenectady, N.Y.

[73] Assignee: **General Electric Company,** Schenectady, N.Y.

[21] Appl. No.: **676,349**

[22] Filed: **Nov. 29, 1984**

[51] Int. Cl.^{4} H01J 17/20; H01J 61/18
[52] U.S. Cl. 313/638; 313/639
[58] Field of Search 313/638, 639, 642, 571, 313/116

[56] **References Cited**

U.S. PATENT DOCUMENTS

3,234,421	2/1966	Reiling	313/25
3,351,798	11/1967	Bauer	313/571
3,398,312	8/1968	Edris et al.	313/642
3,911,308	10/1975	Akutsu et al.	313/184
3,979,624	9/1976	Lui et al.	313/639
4,171,498	10/1979	Fromm et al.	313/116

FOREIGN PATENT DOCUMENTS

1463568 12/1966 France 313/638

OTHER PUBLICATIONS

U.S. Patent Application Ser. No. 454,225, of P. D. Johnson, Filed Dec. 29, 1982, which is basis for English patent application 2133925A published Aug. 1, 1984.

Primary Examiner—Davis L. Willis
Assistant Examiner—William L. Oen
Attorney, Agent, or Firm—Marvin Snyder; James C. Davis, Jr.

[57] **ABSTRACT**

Iodine in excess of sodium-iodide stoichiometry is used in the fill for a sodium iodide arc lamp which also includes sodium iodide and xenon buffer gas. The presence of free sodium near the arc tube walls is eliminated and efficacy is improved.

7 Claims, 1 Drawing Figure

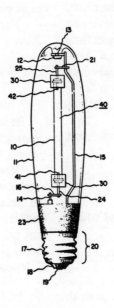

Fig. 5-6. Patent: Sodium Iodide Arc Lamp.

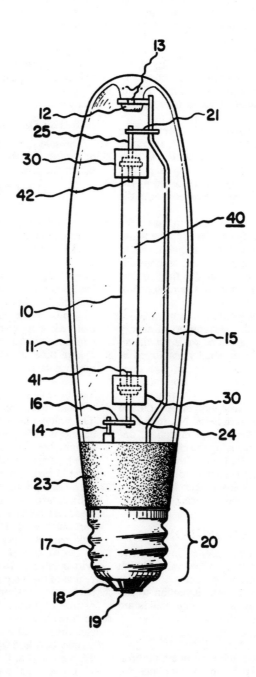

1

HIGH PRESSURE SODIUM IODIDE ARC LAMP WITH EXCESS IODINE

The present invention relates in general to high efficacy high pressure sodium iodide arc discharge lamps and more specifically to the use of excess iodine in a sodium iodide arc discharge lamp.

BACKGROUND OF THE INVENTION

In high intensity arc discharge lamps, the radiated light output is derived from a plasma arc discharge within an arc tube. One form of high intensity discharge lamp that is currently and conventionally employed is the sodium iodide lamp. In such lamps the arc discharge tube includes sodium iodide which is vaporized and dissociated in the plasma arc during lamp operation. However, in the vicinity of the arc tube walls, where the temperature is cooler, sodium remains chemically bound to the iodine limiting the presence of free sodium which absorbs some of the light radiation from the arc discharge.

The self-absorption characteristics of cooler sodium atoms distributed preferentially near the cooler arc tube walls act to limit lamp efficacy. In particular, sodium D-line radiation produced within the hot central plasma region of the arc tube would be readily absorbed by the cooler sodium atoms which would be present near the arc tube walls.

While the use of sodium iodide in the lamp lessens the presence of free sodium near the cooler arc tube walls, the sodium to iodine ratio in this area remains greater than unity. With its smaller atomic mass, sodium diffuses to the arc tube walls more rapidly than iodine. Thus, lamp efficacy is still limited by the presence of free sodium near the arc tube walls.

The high pressure sodium iodide arc lamp requires the use of a buffer gas to limit the transport of energy from the arc discharge to the arc tube walls via chemical reaction. Mercury is conventionally employed as the buffer gas at a high pressure. However, high pressure mercury broadens the sodium D-line radiation toward the red and can tie-up iodine by forming mercury iodide, resulting in more free sodium near the arc tube walls. Copending application Serial No. (676,367), assigned to the assignee of the present invention, discloses xenon buffer gas for improving the efficacy of the high pressure sodium iodide arc lamp. However, even with xenon as the buffer gas, the sodium to iodine ratio in the vicinity of the arc tube walls remains greater than unity (i.e. some free sodium remains) during lamp operation.

OBJECTS OF THE INVENTION

It is a principal object of the present invention to eliminate free sodium near the arc tube walls of high pressure sodium iodide arc discharge lamps.

It is another object of the present invention to improve the efficacy of high pressure sodium iodide arc discharge lamps with xenon buffer gas.

SUMMARY OF THE INVENTION

These and other objects are achieved in a high pressure sodium iodide arc lamp having an arc tube for containing an arc discharge by utilizing an arc tube fill comprising sodium iodide, xenon buffer gas and iodine in sufficient quantity to reduce the partial pressure of sodium at the arc tube walls to zero during lamp opera-

2

tion. The amount of sodium iodide in the lamp provides a sodium pressure in the arc discharge of about 10 to about 100 torr. The excess iodine is provided in an amount which would provide an iodine partial pressure of about 10 to 50 torr in excess of overall sodium-iodine stoichiometry when the lamp is in operation. The iodine in the lamp may be derived from mercury iodide added to the fill.

The present invention further contemplates a high intensity metal halide arc discharge lamp comprising an outer light transmissive envelope, a light transmissive arc discharge tube with electrodes at opposite ends of the arc tube and means to provide electrical connections to the electrodes. A vaporizable discharge medium is disposed within the arc tube, and includes sodium iodide together with xenon buffer gas and an excess of iodine.

The features and advantages of the present invention will become apparent from the following detailed description of the invention when read with the accompanying drawing.

BRIEF DESCRIPTION OF THE DRAWING

The sole drawing FIGURE is a side elevation view of a typical high pressure sodium iodide arc lamp in which the present invention may be embodied.

DETAILED DESCRIPTION OF THE INVENTION

The FIGURE shows a high intensity arc discharge lamp comprising an outer light transmissive envelope 11. This outer envelope preferably comprises a material such as heat resistant glass or silica. The lamp also comprises a light transmissive arc discharge tube 10 which has electrodes disposed internally at opposite ends thereof. Arc discharge tube 10 is typically configured in a cylindrical shape and must be resistant to attack by the materials employed in a gaseous discharge medium 40 contained within the arc tube. In particular, arc discharge tube 10 preferably comprises a refractory ceramic material such as sintered polycrystalline alumina, or may comprise fused quartz. Arc discharge tube 10 may have an internal diameter of about 5 to 20 millimeters and an arc gap of 50 to 150 millimeters, for example. The volume between arc discharge tube 20 and outer envelope 11 is generally evacuated to prevent efficacy robbing heat losses from arc tube 10. Getter material 23 may be disposed on the interior of outer envelope 11 to assist in maintaining vacuum conditions in the volume between arc tube 10 and outer envelope 11.

Structures are shown in the Figure for providing electrical connection and support for arc tube 10. In particular, supporting wire conductors 14 and 15 provide part of a means for connecting the arc tube electrodes 41 and 42 to external connections. Supporting wire conductor 15 extends upward through the vacuum region of the lamp and is preferably welded to a hexagonal bracing washer or ring 13 which is disposed about a dimple 12 provided in the end of an outer envelope 11 to furnish support for arc discharge tube 10. Lateral support wire 21 is preferably spot welded to an arc tube termination lead 25 and to supporting wire conductor 15. Similarly, at the base end of the lamp shown in the Figure, a lateral support 16 is spot welded to supporting wire conductor 14 and a lower arc tube termination 24 so as not only to support arc tube 10 but also to supply electrical current to the electrodes therein. Thus, current through the gaseous discharge medium 40 typically follows a path defined by the following components: supporting wire conductor 14, lower lateral support 16, lower arc tube termination 24, the lower electrode 41 in arc tube 10, gaseous discharge medium 40, the upper electrode 42 in arc tube 10, upper arc tube termination 25, lateral support wire 21, and supporting wire conductor 15. Supporting wire conductors 14 and 15 are separately connected to either of external screw base connection 17 or center exterior contact 19 on edison base 20. Insulating material 18 separates base connection 17 and exterior contact 19.

The lamp shown in the Figure further includes heat shields 30 disposed about the ends of arc tube 10. These heat conserving end shields made of heat insulating material to minimize heat radiation from the ends of arc tube 10, are employed because metal halide lamps require a high temperature to maintain desired vapor pressure of the radiating metal of the lamp fill.

Gaseous discharge medium or fill 40 comprises sodium iodide, xenon buffer gas and an excess of iodine. The amount of sodium iodide in fill 40 will provide a sodium pressure within an arc discharge during lamp operation of about 10 to about 100 torr. Xenon buffer gas is present at a partial pressure of about 100 to about 500 torr at room temperature.

During lamp operation, the vaporized species of fill 40 will adjust their local concentrations so as to provide local thermodynamic equilibrium while balancing diffusion fluxes resulting from concentration gradients. Assuming that the diffusion coefficients of sodium, iodine and sodium iodide in xenon, relative to that of sodium iodide, are 2.53, 1.56, and 1.0, respectively, and assuming equal amounts of sodium and iodine in the vapor phase (i.e. no excess iodine added), the free sodium partial pressure at the arc tube walls during lamp operation will be substantially above zero and the iodine partial pressure at the arc tube walls will be essentially zero. For example, in a lamp with sodium and iodine at sodium-iodine stoichiometry (i.e. all sodium and iodine combined at room temperature), having an arc center temperature of about 4500° K. and an arc tube wall temperature of about 1500° K., and a sodium pressure at the center of the arc of about 52 torr, the free sodium partial pressure at the arc tube walls is about 13 torr and iodine partial pressure at the arc tube walls is zero.

In order to reduce the sodium partial pressure at the arc tube walls to near zero, excess iodine is included in fill 40 at an amount sufficient to provide an iodine partial pressure which is 10 to 50 torr in excess of overall sodium-iodine stoichiometry when the lamp is in operation.

In the present invention, the excess iodine in fill 40 may be derived from mercury iodide added to fill 40. The iodine in the mercury iodide will preferentially combine with free sodium near the arc tube walls during operation of the lamp. The limited amount of mercury added to fill 40 results in a mercury partial pressure too small to cause the problems discussed previously.

The foregoing describes a high pressure sodium iodide arc lamp and fill for such lamp wherein iodine in excess of sodium iodide stoichiometry is added in order to eliminate the presence of free sodium near the arc tube walls during operation of the lamp. The efficacy of the lamp is improved since radiation from the arc discharge is no longer absorbed by free sodium near the arc tube walls.

While preferred embodiments of the present invention have been shown and described herein, it will be obvious that such embodiments are provided by way of example only. Numerous variations, changes and substitutions will occur to those skilled in the art without departing from the invention herein. Accordingly, it is intended that the invention be limited only by the spirit and scope of the appended claims.

What is claimed is:

1. In a high pressure sodium iodide arc lamp having an arc tube for containing an arc discharge, an arc tube fill comprising
sodium iodide;
xenon buffer gas in an amount providing a xenon partial pressure in the range of about 100 torr and higher at room temperature; and
iodine in sufficient quantity to reduce, the partial pressure of sodium at the walls of said arc tube to substantially zero during operation of said lamp.

2. The lamp of claim 1 wherein said sodium iodide is present in sufficient quantity to provide a sodium pressure during operation of said lamp in said arc discharge of about 10 to about 100 torr.

3. The lamp of claim 2 wherein said quantity of iodine equals an amount which provides an iodine partial pressure of 10 to 50 torr in excess of overall sodium-iodine stoichiometry in the presence of an arc within said arc tube.

4. The lamp of claim 1 wherein said iodine is derived from mercury iodide included in said fill.

5. A high intensity arc discharge lamp comprising:
an outer light transmissive envelope;
a light transmissive arc discharge tube situated within said envelope and having electrodes at opposite ends thereof;
means to provide electrical connection to said electrodes;
sodium iodide disposed within said arc tube;
xenon buffer gas disposed within said arc tube in an amount providing a xenon partial pressure in the range of about 100 torr and higher at room temperature; and
iodine disposed within said arc tube in a sufficient quantity to reduce the partial pressure of sodium at the walls of said arc tube to substantially zero during operation of said lamp.

6. The lamp of claim 5 wherein said sodium iodide is present in sufficient quantity to provide a sodium pressure during operation of said lamp in said arc discharge of about 10 to about 100 torr.

7. The lamp of claim 6 wherein said quantity of iodine equals an amount which provides an iodine partial pressure of about 10 to 50 torr in excess of overall sodium-iodine stoichiometry in the presence of an arc within said arc tube.

* * * * *

ent is followed by the word *comprising*. Pay close attention to this word—it it magic. Comprising means "including at least these things." Compare this with *consisting of*. This latter phrase means "including these things and only these things."

Because the arc lamp filler of claim 1 *comprises* the elements (1) sodium iodide; (2) xenon buffer gas; and (3) iodine, it is possible to add something else and still be within the covered territory. Therefore, you could validly write the following dependent claim 2:

> 2. The arc tube fill of claim 1 further comprising chlorine.

Because you have the word *comprising* and you are only adding an additional element (rather than substituting an element for iodine), this revised dependent claim is correct. (Assuming that you have described the use of chlorine in your detailed description!)

Almost all independent patent claims are written with the work *comprising* placed after the preamble, because this results in broader, more valuable claims that competitors cannot easily get around. If *consisting of* is used rather than *comprising*, competitors can get around the patent by merely adding an additional element to the claimed invention. Use of the word *comprising* also permits the valid addition of dependent claims, such as the correct version of claim 2, shown previously.

In some special circumstances, you might be forced to use *consisting of*. This happens primarily with chemical inventions when the addition of anything to the brew results in a non-working product. If you know that certain ingredients and only those ingredients will work, then *consisting of* is the appropriate language. In this circumstance, use of the word *comprising* would result in a claim covering inoperable subject matter. This would violate the requirement that the invention be useful, i.e. the territory covered by the claims.

Claim 5 of the arc lamp patent is an additional independent claim to the arc lamp device itself rather than to just the fill for the lamp. Read the preamble of both claims carefully in order to see the difference. It is good patent claiming technique to use more that one independent claim and more than one approach toward defining the invention. Claim 1 defines the invention as a fill for arc lamps, whereas claim 5 approaches the invention as a new type of arc lamp . . . one containing a different type of fill. Claiming the invention in different ways provides back-up protection should one of the approaches prove invalid.

USE OF GENERAL LANGUAGE IN CLAIMS

Look carefully at the structure of claim 5. The preamble is:

> 5. A high intensity arc discharge lamp comprising:

After the preamble, there are 5 sub-blocks of text. Each of these blocks describes an element of the invention. In this case, the elements are

> an outer transmissive envelope
> a light transmissive arc discharge tube
> means to provide electrical connection
> sodium iodide
> xenon buffer gas, and
> iodine . . .

Notice that very general language is used where possible. Why does the claim refer to an "outer transmissive envelope" rather than to a "glass bulb"? Wouldn't it be simpler to just say "glass"? Yes, it would be simpler. However, if just the word glass was used, competitors could avoid getting into trouble by using clear quartz or some other clear substance in place of glass. Furthermore, if the phrase "clear envelope" was used, competitors could get around the patent by using tinted glass or by making the glass cloudy. But there is no obvious way to avoid using a "transmissive envelope." In order for the lamp to work, it must transmit light from the inside to the outside. This, by definition, is what a transmissive envelope does. By closely linking the definition of an element to the function that element must perform the writer of this claim arrived at language that both clearly describes the element and makes the patent difficult to avoid.

Use "means for" language to broaden claims. The third element of claim 5, "means to provide electrical connection to said electrodes . . . " illustrates a very clear way of putting the purpose of an element into the definition of the element. This language covers any known way for getting electricity to electrodes whether it is by soldered wires, spring contacts, screw-on clamps, or whatever. Because everything is covered by his language, it will be hard for a competitor to substitute something that will successfully avoid the patent.

The patent statute specifically authorizes the use of "means for . . . doing something " language for describing a *well known* element of an invention. This is justified in the case of the arc lamp patent, because ways of connecting to electrodes are common. However, before you use "means for" language, be sure that the means is well known among ordinary technicians. When you write a patent application for a flying saucer, you won't get away with language like "means for neutralizing the gravitational field in the vicinity of said vehicle . . . " Since a means for neutralizing gravity is not well known, you must use language that defines how neutralization of gravity is to be accomplished.

Claims for mixtures of ingredients can be written by, more or less, listing the ingredients as elements. Claim 1 of Fig. 5-6, for example, merely lists what materials are to be put inside an arc lamp. However, a mere listing of elements is not adequate for other types of inventions.

DO NOT WRITE "BOX OF PARTS" CLAIMS

When writing a claim for a manufacture or machine, you must do more than simply list the elements: you must indicate how each element attaches to or works with other elements to accomplish a desired result. A machine is more than a box of parts. It is a box of parts that has been assembled into a coordinated whole that can perform a function. A claim for a machine must reflect this.

Claim 1 of the arc lamp patent covers only a composition of matter. In this case, a mere listing of ingredients can be acceptable. But claim 5 goes farther: it is directed to a complete arc lamp device, as a machine or manufacture. Therefore, claim 5 must contain language that explains how the various parts are connected to each other. For example, the light transmissive arc discharge tube is said to be "situated within said envelope . . . " The means to provide electrical connection is "to said electrodes." The sodium iodide is " . . . disposed within said tube." Thus, each element of claim 5 is listed and then followed with a short phrase explaining how that element is connected to something else.

THE LAND AND BRIDGE TEST

Here is a good way to check the structure of machine or manufacture claims to be sure that you have included adequate language explaining the connections among the various parts. On a piece of paper, represent each element of the claim as an island. In the case of the arc lamp in claim 5, there will be 6 islands, as shown in Fig. 5-7. Then, represent the connecting words in your claim as bridges between the islands. Figure 5-7 illustrates a completed diagram with bridges.

If your claim, as diagrammed, contains an island that is not connected to at least one other island by a bridge, you have left out important connecting language. Rewrite the claim to include additional connecting words!

Island and bridge diagrams can also be used to check dependent claims. First, diagram the independent claim. Then ask this question: What must be done to this diagram in order to arrive at a diagram representing what is described by the total of the independent claim and the dependent claim? If you must remove an island and substitute another island, i.e. take one element out and put in another element, then you do not have a valid dependent claim. This problem is illustrated in Fig. 5-8. However, if the dependent claim merely adds another bridge and island to the diagram, as in Fig. 5-9, it is good. If the dependent claim merely modifies or qualifies the island in some way, as in Fig. 5-10, it is also good.

Another example of a valid dependent claim that merely modifies one of the elements stated in the independent claim, is seen in claim 6 of Fig. 5-6. Claim 6 merely takes the element "iodine" of claim 5 and specifies something more about the iodine—that it is "present in sufficient quantity to provide a sodium pressure . . . of about 10 to about 100 torr."

CLAIMS FOR IMPROVEMENT INVENTIONS

If the invention is an improvement, another form of claim is possible. It is called the Jepson format, because Jepson's patent lawyer was among the first to try it. Jepson claims have this format:

> "A [something old] . . . wherein the improvement comprises . . . [describe improvement].
>
> Example: "An improved arc lamp fill of the type comprising iodine, sodium iodide, and xenon, where in the improvement comprises the addition of radon gas."

Since the Jepson format claim focuses clearly on the very item you have improved, the Patent Office recommends its use. However, courts tend to interpret Jepson claims narrowly. They say that by using a Jepson claim, the inventor has admitted everything described before the word *comprising* to be old. If all of this really was not old, the inventor can be deprived of part of his invention.

Due to this unfavorable attitude toward Jepson claims, you should probably not put claims into Jepson format unless you are absolutely certain about what is new and what is old.

METHOD CLAIMS FOR SOFTWARE

Claims to methods or processes require another approach to claim writing. In a method or process, the elements are not physical things such as the parts of a machine. The elements of a method or process are steps for doing something. Instead of nouns such as "xenon gas" or "transmissive envelope," there are verb forms such as "cracking," "heating," or "calculating." Method claims are especially useful for writing patents that protect software. Because a computer program is a sequence of steps to be taken by a computer in arriving at some desired output, it is natural to claim these steps in a patent as the elements of a method claim.

Good examples of method claims are seen in the patent of Fig. 5-5, which covers a new way to use a computer for word processing. The preamble of claim 1 is typical of method claims:

> "1. A method of . . . [doing something] . . . said method comprising the steps of."
>
> In Fig. 5-5, the [doing something] is "creating computer text."

Notice that the word comprising can be used for steps as well as for physical elements (such as machine parts). Use of this word prevents others from avoiding the patent by merely adding additional steps to the patent.

In addition to identifying Fig. 5-5 claim 1 as a method claim, the preamble also introduces and places in context items that must be mentioned in order to describe the steps of the method. The preamble introduces the words *memory*, *screen*, and *means for selecting positions* so that these terms do not pop out of nowhere when the steps of the method are described. Since the word *memory* is introduced

Fig. 5-7. Here is an island and bridge diagram for the above claim language, claim 5 of U.S. Patent 4,605,881. Each element of the claim is shown as an island. Language showing how the islands are connected into a whole invention is illustrated with bridges. Because a bridge exists to connect each island to at least one other island, the language of claim 5 passes the island and bridge test.

Claim Language:

5. A high intensity arc discharge lamp comprising an outer light transmissive envelope;

a light transmissive arc discharge tube situated within said envelope and having electrodes at opposite ends thereof;

means to provide electrical connection to said electrodes,

sodium iodide disposed within said arc tube,

xenon buffer gas disposed within said arc tube in an amount providing a xenon partial pressure in the range of about 100 torr and higher at room temperature; and

iodine disposed within said arc tube in a sufficient quantity to reduce the partial pressure of sodium at the walls of said arc tube to substantially zero during operation of said lamp.

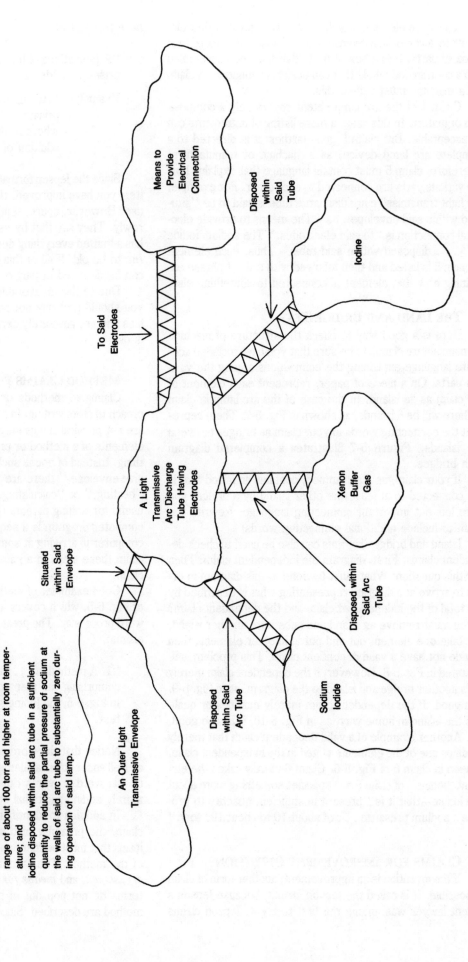

Independent Claim on which dependent claim depends:

5. A high intensity arc discharge lamp comprising

an outer light transmissive envelope;

a light transmissive arc discharge tube situated within said envelope and having electrodes at opposite ends thereof;

means to provide electrical connection to said electrodes,

sodium iodide disposed within said arc tube,

xenon buffer gas disposed within said arc tube in an amount providing a xenon partial pressure in the range of about 100 torr and higher at room temperature; and

iodine disposed within said arc tube in a sufficient quantity to reduce the partial pressure of sodium at the walls of said arc tube to substantially zero during operation of said lamp.

The *improper* dependent claim language:

8. The lamp of claim 5 wherein chlorine is substituted for iodine.

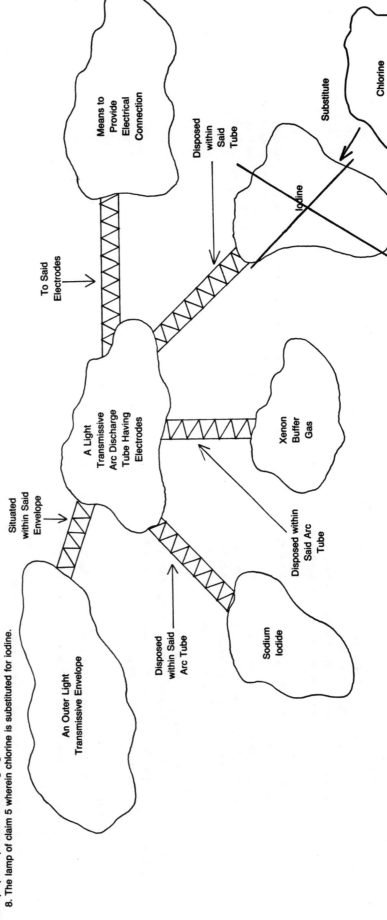

Fig. 5-8. Example of an improper dependent claim. To arrive at the coverage of claim 6 from that of claim 5, the island representing iodine must be removed and an island representing chlorine put in its place. But islands may only be added or modified, never taken out or substituted. Thus, Claim 6 is invalid.

Claim Language: Independent Claim

5. A high intensity arc discharge lamp comprising
an outer light transmissive envelope;

a light transmissive arc discharge tube situated within
said envelope and having electrodes at opposite
ends thereof;

means to provide electrical connection to said electrodes,

sodium iodide disposed within said arc tube,

xenon buffer gas disposed within said arc tube in an
amount providing a xenon partial pressure in the
range of about 100 torr and higher at room temperature; and

iodine disposed within said arc tube in a sufficient
quantity to reduce the partial pressure of sodium at
the walls of said-arc tube to substantially zero during operation of said lamp.

Dependent claim:
9. The lamp of claim 5 additionally comprising radon gas disposed within said arc tube.

Fig. 5-9. Dependent claim adding an additional element. Because the dependent claim merely adds an additional element and is properly associated with connecting language (represented as an additional bridge), this dependent claim is good.

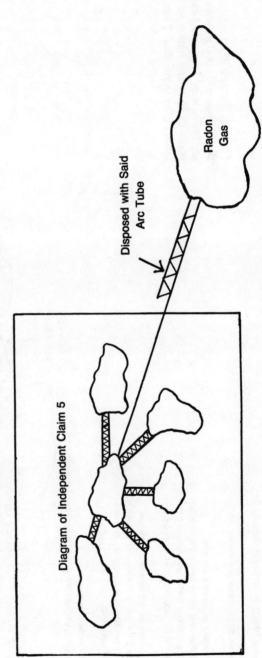

Diagram of Independent Claim 5

Disposed with Said Arc Tube

Radon Gas

Additional bridge of connecting language and additional element of invention, indicated as island.

Independent Claim:

5. A high intensity arc discharge lamp comprising

an outer light transmissive envelope;

a light transmissive arc discharge tube situated within said envelope and having electrodes at opposite ends thereof;

means to provide electrical connection to said electrodes,

sodium iodide disposed within said arc tube,

xenon buffer gas disposed within said arc tube in an amount providing a xenon partial pressure in the range of about 100 torr and higher at room temperature; and

iodine disposed within said arc tube in a sufficient quantity to reduce the partial pressure of sodium at the walls of said arc tube to substantially zero during operation of said lamp.

First Dependent Claim:

10. The lamp of claim 5 additionally comprising radon gas disposed within said arc tube.

Dependent Claim Dependent from a Previous Dependent Claim:

11. The lamp of claim 10 wherein said radon gas is present in an amount providing a pressure in excess of 5 torr at room temperature.

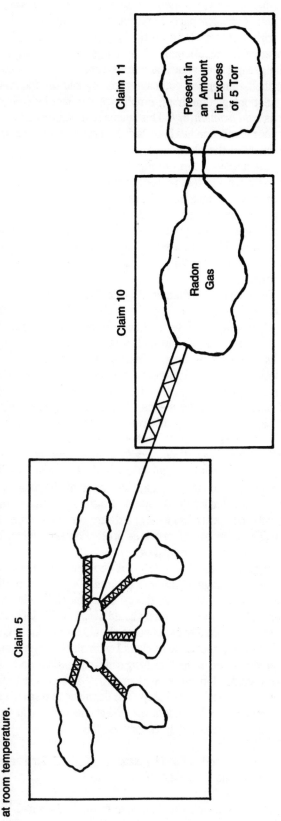

Fig. 5-10. Dependent claim dependent on another dependent claim. Because dependent claim 11 merely adds or qualifies an element present in the claim on which it depends, it is valid. This example assumes radon was properly mentioned and discussed in the detailed description portion of the patent application.

Claim 5

Claim 10

Radon Gas

Claim 11

Present in an Amount in Excess of 5 Torr

in the preamble, it may be referred to later in the claim as *said memory*, as is seen in element (a) of claim 1.

After you write a method claim, inspect every step for mention of physical parts or objects that are used in connection with that step. Be sure that you have properly laid the foundation for mentioning these things by describing them in the preamble of your claim or as aspects of a previously described element. If you have properly laid the foundation for mentioning something, emphasize this fact by referring to it as *said [something]*. This language emphasizes the fact that your foundation has been laid. You have "said" this word previously in the claim.

The steps of claim 1 of Fig. 5-5 are listed in outline form with identifying letters in front of each step or element. The use of outline format and letters is not required, but it does make the claim easier to read. Notice that portions of text describing each element are set apart from one another with semicolons. After the semicolon of the second to last element, insert the word *and* so that the entire claim reads like a sentence.

The elements of method claims in Fig. 5-5 are verb forms ending with—ing: "storing," "displaying," and "identifying." These words describe general functions performed by all computers. "Storing in said memory " can apply to semiconductor memory, floppy disk, or optical disk. Likewise, "displaying on . . . said screen" covers CRT displays as well as liquid crystal displays. Furthermore, "identifying . . . with said position selecting means" can be accomplished with a light pen, a mouse, a touch screen, or a keyboard. If the claim had used the term "CRT" rather than "screen," or "floppy disk" rather than "memory," competitors could have avoided the patent by simply using LCD's or an optical disk. By carefully choosing words, the writer of this patent obtained protection for the computer program not limited to any particular computer language, hardware, or even screen layouts. Thus the patent provides protection far beyond that available under copyright law.

Figure 5-5 demonstrates that dependent claims can be used to advantage in method patents. Dependent method claims must do one of two things: They must (1) add an additional step to the previously listed steps or (2) define how one of the previously listed steps should be modified. Claim 2 covers the same steps as claim 1 with step (a) being modified so that some of the "said linguistic expressions each comprise a single alphanumeric character." In contrast to this, claim 6 adds an additional step. The linguistic pattern of claim 6 is standard for dependent method claims which add an additional step. It is highly recommended:

> "The method defined by claim _____ further comprising the . . . step of . . ."

Do not put ordering language in method claims. When writing method claims, do not inadvertently use ordering words such as "the first step is," or "the next step is,"
or "then" or "finally," or any other expression that suggests the steps must be performed in some particular order. If a method claim has words such as these, a competitor will be able to avoid the patent by merely changing the order of steps. If a claim does not contain ordering words, the patent will cover any method which includes all of the listed steps, regardless of order. Thus, you should include ordering words only if a certain order is essential and not apparent from common sense. Example: it is discovered that an old method works much better if step (c) is done before step (a). Because no one previously realized the special benefit to be gained from this particular order, it is possible to patent an improved method wherein step (c) is performed before step (a). However, because the invention is not the entire method but only an improved order, the claim must expressly state that step (c) is to be done before step (a).

NEW USE CLAIMS

If the invention is a new use for something old, such as a previously known chemical compound, you can use a claim with the following form:

What I claim is:

1. The use of plutonium nitrate powder as an insecticide.

HOW TO WRITE CLAIMS: AN EASIER WAY

Writing patent claims from scratch is not easy, Is there a simpler way? The answer is often yes. You can simplify the task by obtaining copies of other people's patents in the same field of technology. If you have designed a new type of light bulb, search for all previous patents on light bulbs. Study the form and wording of the claims. Pay special attention to patents owned by, say, General Electric Company. The person who wrote these could well be a patent attorney who has devoted his life to writing claims for light bulbs. Only such a person could be warped enough to refer to the glass of a light bulb as a "light transmissive envelope." As you read numerous light bulb patents, you will develop a mental thesaurus of technical terms useful for writing claims in your favorite area of technology.

Notice the approach of the claims as well as the words used. Was the light bulb claimed as a device or as a method for producing light from electricity? Was the whole light bulb claimed, or just the filament material? Did the attorney use a Jepson/improvement claim format? Why was the preamble worded the way it was?

As you read these patents, ask yourself this question: How could a competitor get around this claim language? What alternative materials or ways of constructing the light bulb does this language cover? By asking yourself these questions, you will come to appreciate the careful thought that goes into claim language. You will soon be incorporating some of this craftsmanship into your own writing.

HOW TO WRITE CLAIMS: THE EASIEST WAY OF ALL

You must include at least one claim with your patent application in order to have a valid filing. However, if the patent examiner rejects your claims, and if you are filing a patent application for your own invention without an attorney, you can request the patent examiner to write a suitable claim for your application. If the patent examiner is persuaded that you have presented patentable material in your detailed description, he will write a patent claim which he thinks covers this material and allow the patent. Naturally, the patent examiner may not write a claim that is the best possible claim. Patent examiners are notoriously overworked and have things to worry about aside from helping you write your claims. However, as a last resort, this technique can be useful.

CHAPTER 6

Step 2: Send Application et al To The Patent Office

The package of papers to be filed at the Patent Office in Washington, D.C. to begin the application process includes the following:

(1) Patent Application
(2) Rough Drawings
(3) Abstract of the Disclosure
(4) Declaration for a Patent Application
(5) Verification of Small Entity Status
(6) Filing Fee
(7) Patent Application Transmittal Letter
(8) A Stamped, Self-addressed Reply Postcard

Ideally, all of these items should be enclosed in a single envelope and mailed, first class, to the following address:

Commissioner of Patent and Trademarks
Washington, D.C. 20231

Chapter 5 discusses the preparation of item (1). This chapter discusses preparation of the remaining items and assembly of these items into a proper package, ready for mailing.

(2) PREPARING THE ROUGH DRAWINGS

The patent statute requires that drawings accompany patent applications when these will be of use in illustrating the invention. Some inventions, such as a new formula for shampoo, for example, are not readily presented in a drawing. For these, a mere list of ingredients in the detailed description and claims will suffice. However, for the usual electronic and mechanical inventions, a drawing will be required.

Drawings must be more than mere pictures of the invention. They must be carefully coordinated with the text of the detailed description. To achieve this coordination, the detailed description must refer to each drawing and, must discuss and describe, using reference numbers, the parts shown within each drawing.

Before a patent can be issued, the applicant must (usually) submit formal line drawings in india ink. These are rather difficult to prepare and are usually done by professionals hired by the inventor. The patent office can supply a list of bonded draftsmen who are qualified to prepare formal drawings, upon request.

However, formal drawings need not be filed until a patent has been approved for issuance. At the time of filing, only rough, pencil sketches hand-drawn by the inventor are required. Indeed, it is wise *not* to get formal drawings done until the patent examiner has seen and approved the rough drawings. This avoids the need for expensive corrections.

Rough drawings should have all details and reference numbers that are to appear eventually in the formal drawings. They can be done on ordinary bond paper. Each rough drawing should be labeled at the bottom with a figure number. The word figure should be abbreviated, as in "Fig. 1." If the same part is shown in two different figures, it should be given the same reference number in each. It is best to put each figure on a separate page.

(3) THE ABSTRACT OF THE DISCLOSURE

The abstract is a 50-to 200-word summary of what is disclosed in the patent application. It should be typed on a separate sheet of paper.

At the top of the separate paper, place the words: "Abstract of the Disclosure." Then, in a single paragraph, briefly describe the invention. Do not refer to drawings in this summary.

(4) THE DECLARATION

The *declaration* is the inventor's certification, under penalty of perjury, that he is indeed the inventor of the thing shown in the application and that, to the best of the inventors knowledge, other requirements for patentability are present. All of these requirements are set forth in the declaration itself. Figure 6-1 is a sample declaration. A blank declaration form, suitable for photocopying, is included in Appendix A.

Every patent application must be supported by a declaration. Normally, the declaration form is signed by the inventor and included with the original filing of the application with the Patent Office.

If there is more than one inventor, it is normally required that all inventors sign. In abnormal situations such as when one inventor cannot be found or when he refuses to participate in signing the declaration, the patent application can still be filed. However, the Patent Office will require that an affidavit be filed explaining the circumstances. If an inventor is deceased, it is possible for a representative of the inventor's estate to execute an declaration with suitably modified wording.

The declaration does not have to be filed with the patent application. Thus, if desirable for reasons of convenience, a patent application can be filed to obtain an early filing date without the need for the inventor to physically sign a document. For this convenience, however, the Patent Office demands an additional fee.

The declaration does not have to be witnessed or notarized, provided it contains language to the effect that the declaration is made under penalty of perjury. The declaration form seen in Fig. 6-1 is suitable for use without a notary. it is suitable for use with up to two inventors.

Blank A on the form must be filled in with the title of the invention as it appears on the first page of the patent application.

Blank B is marked if the declaration form is filed in the Patent Office at the same time as the patent application. For declarations filed later, blank C is marked.

If blank C is marked, blank C1 should be filled in with the date of filing, C2 should be filled in with the serial number assigned to the application from the Patent Office, and blank C3 should be filled with the date of any amendment that accompanied the original filing or was later filed. In the normal case, there is no amendment and blanks C1 through C3 are not filled.

Paragraph D of the declaration states that the inventors have actually read the patent application and that they understand it. This means that the declaration form should never be signed unless the inventors have had physical access to a copy of the patent application document. Thus, mailing a copy of the declaration form to an inventor for signing without also including a copy of the patent application is improper.

In paragraph E of the declaration, the inventors promise to tell the Patent Office if they ever discover any problems with the patent application. The following example illustrates the significance of paragraph E.

Example: After a patent application is filed, an inventor runs across a magazine article which suggests that the invention was previously made by someone else. Can the inventor merely ignore this article and hope that the patent examiner does not find it?

Answer: No! The law is that the inventor owes a duty of good faith and candor to the Patent Office. Paragraph E of the inventor's declaration binds him to tell the examiner about any problems, even if the inventor was not aware of these problems at the time the declaration was signed. The inventor must write a letter to the Patent Office to inform the examiner of the magazine article. Failure to do so can result in invalidity of the patent and very serious penalties for violation of various laws.

Paragraph F has to do with previous foreign patent filings for the same invention. Because the law requires that U.S. inventors file in the U.S. Patent Office at least 6 months before filing abroad, there should be no reason to fill in any of the blanks in this paragraph.

Paragraph G has to do with previous U.S. applications. If the patent application is for an invention that was also described in a previous patent application, the serial number and filing date of the previous application should be filled in.

Paragraph H is the language that allows the declaration to be completed without a notary or witnesses. The full name of the inventor should be filled in blank H-1. Do not use abbreviations or initials. The full signature of the inventor, without abbreviations or initials, should be inserted at blank H-2. The date is required at blank H-3. *Critical:* The declaration will not be valid if it is not dated or if the typed name in H-1 is not exactly like the signed name in H-2. For example, it is invalid to type the name as ''George William Smith'' but sign as ''George Smith.'' This is an annoying technicality, but that's the way it is.

The blanks in paragraph I must be filled in with the residence address, citizenship, and post office address of the inventor.

The declaration form has duplicate paragraphs for the signatures and residence addresses, etc. for two inventors. If there are more than two inventors, additional form paragraphs and blanks will have to be added.

Important: At the time that each inventor signs the oath, he or she must have an actual copy of the patent application. Furthermore, NO CHANGES WHATSOEVER TO THE APPLICATION MAY BE MADE AFTER THE OATH IS SIGNED.

Who must sign the oath?

The declaration must be signed by the inventor or, if there is more than one inventor, by all of the inventors. Notice that *only* inventors may sign. Persons who merely help

OMB No. 0651-0011 (12/31/86)

DECLARATION FOR PATENT APPLICATION   Docket No. _____

As a below named inventor, I hereby declare that:

My residence, post office address and citizenship are as stated below next to my name.

I believe I am the original, first and sole inventor (if only one name is listed below) or an original, first and joint inventor (if plural names are listed below) of the subject matter which is claimed and for which a patent is sought on the invention entitled
(A) _____Rust-Proof Mousetrap_____, the specification of which

(B) (check one) ☒ is attached hereto.
(C) ☐ was filed on **(C1)** _____ as
Application Serial No. **(C2)** _____
and was amended on **(C3)** _____ (if applicable).

(D) I hereby state that I have reviewed and understand the contents of the above identified specification, including the claims, as amended by any amendment referred to above.

(E) I acknowledge the duty to disclose information which is material to the examination of this application in accordance with Title 37, Code of Federal Regulations, §1.56(a).

(F) I hereby claim foreign priority benefits under Title 35, United States Code, §119 of any foreign application(s) for patent or inventor's certificate listed below and have also identified below any foreign application for patent or inventor's certificate having a filing date before that of the application on which priority is claimed:

Prior Foreign Application(s)   Priority Claimed

(Number)	(Country)	(Day/Month/Year Filed)	Yes	No
(Number)	(Country)	(Day/Month/Year Filed)	Yes	No
(Number)	(Country)	(Day/Month/Year Filed)	Yes	No

(G) I hereby claim the benefit under Title 35, United States Code, §120 of any United States application(s) listed below and, insofar as the subject matter of each of the claims of this application is not disclosed in the prior United States application in the manner provided by the first paragraph of Title 35, United States Code, §112, I acknowledge the duty to disclose material information as defined in Title 37, Code of Federal Regulations, §1.56(a) which occurred between the filing date of the prior application and the national or PCT international filing date of this application:

(Application Serial No.)	(Filing Date)	(Status—patented, pending, abandoned)
(Application Serial No.)	(Filing Date)	(Status—patented, pending, abandoned)

I hereby appoint the following attorney(s) and/or agent(s) to prosecute this application and to transact all business in the Patent and Trademark Office connected therewith:

Address all telephone calls to _John W. Inventor____ at telephone no. _(202) 555-2300_.
Address all correspondence to
John W. Inventor
P.O. Box 345
White Plains, NY 34545

(H) I hereby declare that all statements made herein of my own knowledge are true and that all statements made on information and belief are believed to be true; and further that these statements were made with the knowledge that willful false statements and the like so made are punishable by fine or imprisonment, or both, under Section 1001 of Title 18 of the United States Code and that such willful false statements may jeopardize the validity of the application or any patent issued thereon.

(H1) Full name of sole or first inventor _John William Inventor_
(H2) Inventor's signature _____ Date _1/2/92_ **(H3)**
(I) Residence _800 Main St., White Plains, NY 34545_ Citizenship _United States_
Post Office Address _P.O. Box 345, White Plains, NY 34545_

Full name of second joint inventor, if any _George Arthur Collaborator_
Second Inventor's signature _____ Date _1/2/92_
Residence _2525 Union St., Teaneck, NJ 44765_ Citizenship _United States_
Post Office Address _2525 Union St., Teaneck, NJ 44765_

(Supply similar information and signature for third and subsequent joint inventors.)

Form PTO-FB-A110 (8-83)

Fig. 6-1. Sample Declaration for Patent Application.

with the financing, routine construction, or marketing of an invention are not co-inventors. To be an inventor, a person must contribute to the idea behind the invention. In some sense, that idea must be his or her idea. If there is more than one inventor, then all of the co-inventors must have somehow mixed together their thought processes and come up with a new concept.

Example: Smith and Jones have a discussion about how electricity can be used to create light. Smith suggests that electric current be run through carbonized cotton fiber. Jones points out that the fiber will burn unless it is in a vacuum. Smith suggests that the vacuum be created in a glass bulb. Smith and Jones ask Johnson, a technician, to build a glass bulb with a carbonized fiber in accordance with their plans. Hughes, a banker, invests money in the project in order to pay Johnson's salary. Who must sign the oath?

Answer: On these facts, only Smith and Jones are inventors. Therefore, only Smith and Jones may sign the oath. Hughes can get a part of the invention, but only if a contract is signed that assigns Hughes part of the patent rights as assignee.

(5) PREPARING THE VERIFICATION OF SMALL ENTITY STATUS

Patent Office fees are one-half the normal rate for "small entities," that is, for individual inventors and businesses with fewer than 500 employees. To qualify for small entity rates, a paper verifying small entity status must be filed. A sample verification appears in Fig. 6-2. Appendix A contains a blank form suitable for photocopying.

(6) CALCULATING THE PATENT OFFICE FILING FEES

The basic patent application filing fee is $170 for patent applications filed by individual inventors or businesses, provided a verification of small entity status has been filed. This fee should be paid with a check or money order accompanying the patent application.

The basic $170 filing fee covers an application with up to 3 independent claims and 20 total claims. For each independent claim in excess of three, there is an additional fee of $17. For each claim of any type in excess of 20 total, there is an additional fee of $6.

Example: Suppose a patent application contains four independent claims and 17 dependent claims. What is the fee, assuming that a paper claiming small entity status has been filed?

Answer: There are 21 total claims. Therefore, in addition to the basic $170 fee, there is a $6 fee for one claim in excess of 20. Furthermore, there is one independent claim in excess of the three allowed. This costs an additional $17. Therefore, the total fee is:

$$\$170 + \$6 + \$17 = \$193$$

Notice that the addition of one independent claim resulted in both a $6 fee for a claim over 20 and $17 fee for an independent claim in excess of three. This is how the Patent Office makes money.

You can file a patent application without paying the entire fee or any fee at all. The Patent Office will bill you. However, this will result in an additional surcharge of $55.

If more claims are added to a patent application by amendment, you must pay additional fees if the resulting application contains claims in excess of those allowed under the basic fee. Additional fees can be avoided if the response to the office action directs the examiner to cancel a specific old claim at the same time as a new one is added.

A complete list of Patent Office fees is given in Appendix C.

TRANSMITTAL LETTER

The patent application transmittal letter is not a required document, however it is a useful checklist. Figure 6-3 is a sample transmittal letter form that has been marked with reference letters for purposes of this discussion. A blank form suitable for photocopying is presented in Appendix A.

The following paragraphs, marked with capital letters, refer to corresponding letters marked on Fig. 6-3.

A.   Fill in inventor's names.

B.   Fill in invention title as it appears on the application.

C.   Fill in number of pages of drawings.

D.   This is used if a written assignment of patent rights to a third party has been made prior to filing—otherwise leave blank.

E.   This is used if a foreign application has been previously filed—otherwise leave blank.

F.   Used only if an attorney is filing the case and he wishes to allow his assistant to also deal with the patent office.

G.   Check box if verification of small entity status form is included with patent application.

H.   Use these boxes for calculating the proper filing fee.

    ✍ For small entity rates, calculate number of claims, whether dependent or independent, in excess of 20 and enter this in the upper "NO. EXTRA" box.

    ✍ Calculate number of independent claims in excess of 3 and enter in lower "NO. EXTRA" box.

    ✍ Count number of multiple dependent claims present and write number the "MULTIPLE DEPENDENT CLAIM PRESENT" box. A multiple dependent claim is a claim that depends on more than one previous claim.

Example: Assume that claims 1 and 2 are independent claims for light bulb. An ordinary dependent claim refers to claim 1 or claim 2, but not both as in

"3. The light bulb of claim 2 further comprising . . . "

In contrast to this, a multiple dependent claim refers to more than one previous claim, as in

"The light bulb of claim 1 or claim 2 further comprising . . . "

Multiple dependent claims are used to eliminate the need for writing several dependent claims. However, they cause extra work for the patent office, so a $55 surcharge for each such claim is charged.

    ✍ Multiply the number of extra claims in each category by the rate shown in the "SMALL ENTITY" box, and figure the total.

I.   Check this box if you have set up a pre-paid deposit account with the Patent Office and wish to have filing fees deducted from this. Otherwise leave blank.

J.   Check this box and include a photocopy of the transmittal letter.

K.   Fill in the amount of the enclosed check or money order.

L.   Used only in connection with deposit accounts.

M.   Date and sign form.

OPTIONAL: Enter you personal filing system number for this patent application in the "ATTORNEY'S DOCKET NO." box if desired. Otherwise, leave blank.

(8) REPLY POSTCARD

To verify that your patent application actually arrives at the Patent Office in the mail, it is customary to include with each patent filing a self-addressed, stamped rely postcard. Address the card to your mailing address. On the other side print the following:

    ✍ TITLE OF INVENTION: (Fill in Title)
    ✍ DATE OF FILING: (Leave blank for Patent Office to fill in.)
    ✍ SERIAL NUMBER OF APPLICATION: (Leave blank)

When the postcard comes back, be sure to record the application serial number in a safe place. This allows you to prove your filing in the event the Patent Office misplaces your application.

Applicant or Patentee: ___John William Inventor,et al_____ Attorney's
Serial or Patent No.: _(fill in if patent has_____ Docket No.: _____
Filed or Issued: _____already been filed)
For: __Rust-Proof Mousetrap_____

VERIFIED STATEMENT (DECLARATION) CLAIMING SMALL ENTITY
STATUS (37 CFR 1.9 (f) and 1.27 (b)) — INDEPENDENT INVENTOR

As a below named inventor, I hereby declare that I qualify as an independent inventor as defined in 37 CFR 1.9 (c) for purposes of paying reduced fees under section 41 (a) and (b) of Title 35, United States Code, to the Patent and Trademark Office with regard to the invention entitled _Rust-Proof Mousetrap_____ described in

    [×] the specification filed herewith
    [ ] application serial no. _____ , filed _____ .
    [ ] patent no. _____ , issued _____ .

I have not assigned, granted, conveyed or licensed and am under no obligation under contract or law to assign, grant, convey or license, any rights in the invention to any person who could not be classified as an independent inventor under 37 CFR 1.9 (c) if that person had made the invention, or to any concern which would not qualify as a small business concern under 37 CFR 1.9 (d) or a nonprofit organization under 37 CFR 1.9 (e).

Each person, concern or organization to which I have assigned, granted, conveyed, or licensed or am under an obligation under contract or law to assign, grant, convey, or license any rights in the invention is listed below:

    [×] no such person, concern, or organization
    [ ] persons, concerns or organizations listed below*

        *NOTE: Separate verified statements are required from each named person, concern or organization having rights to the invention averring to their status as small entities. (37 CFR 1.27)

FULL NAME _____
ADDRESS _____
        [ ] INDIVIDUAL   [ ] SMALL BUSINESS CONCERN   [ ] NONPROFIT ORGANIZATION

FULL NAME _____
ADDRESS _____
        [ ] INDIVIDUAL   [ ] SMALL BUSINESS CONCERN   [ ] NONPROFIT ORGANIZATION

FULL NAME _____
ADDRESS _____
        [ ] INDIVIDUAL   [ ] SMALL BUSINESS CONCERN   [ ] NONPROFIT ORGANIZATION

I acknowledge the duty to file, in this application or patent, notification of any change in status resulting in loss of entitlement to small entity status prior to paying, or at the time of paying, the earliest of the issue fee or any maintenance fee due after the date on which status as a small entity is no longer appropriate. (37 CFR 1.28 (b))

I hereby declare that all statements made herein of my own knowledge are true and that all statements made on information and belief are believed to be true; and further that these statements were made with the knowledge that willful false statements and the like so made are punishable by fine or imprisonment, or both, under section 1001 of Title 18 of the United States Code, and that such willful false statements may jeopardize the validity of the application, any patent issuing thereon, or any patent to which this verified statement is directed.

John William Inventor   George Arthur Collaborator
NAME OF INVENTOR   **NAME OF INVENTOR**   **NAME OF INVENTOR**

Signature of Inventor   **Signature of Inventor**   **Signature of Inventor**

  1/2/92   1/2/92
Date   **Date**   **Date**

Form PTO-FB-A410 (8-83)

Fig. 6-2. Sample Verification of Small Entity Status.

Fig. 6-3. Patent Office Transmittal Letter.

OMB No. 0651-0011 (12/31/86)

PATENT APPLICATION TRANSMITTAL LETTER	ATTORNEY'S DOCKET NO. (OPTIONAL)

(A) TO THE COMMISSIONER OF PATENTS AND TRADEMARKS:

Transmitted herewith for filing is the patent application of ___John William Inventor___ and

___George Arthur Collaborator___

(B) for ___Rust-Proof Mousetrap___

Enclosed are:

(C) ☒ ___2___ sheets of drawing.

(D) ☐ an assignment of the invention to ___

(E) ☐ a certified copy of a ___ application.

(F) ☐ associate power of attorney.

(G) ☐ verified statement to establish small entity status under 37 CFR 1.9 and 1.27. ———

(H)

CLAIMS AS FILED			SMALL ENTITY			OTHER THAN A SMALL ENTITY	
FOR.	NO. FILED	NO. EXTRA	RATE	FEE	OR	RATE	FEE
BASIC FEE				$170	OR		$340
TOTAL CLAIMS	35	-20-	•15	x$6=	$ 90	OR	x$12=	$
INDEP CLAIMS	5	-3-	•2	x$17=	$ 34	OR	x$34=	$
MULTIPLE DEPENDENT CLAIM PRESENT	none		+$55=	$ 0	OR	+$110=	$
* If the difference in col. 1 is less than zero, enter "0" in col. 2			TOTAL	$ 294	OR	TOTAL	$

(I) ☐ Please charge my Deposit Account No. ___ in the amount of $ ___.

(J) ☒ A duplicate copy of this sheet is enclosed.

(K) ☐ A check in the amount of $ ___294.00___ to cover the filing fee is enclosed.

(L) ☐ The Commissioner is hereby authorized to charge payment of the following fees associated with this communication or credit any overpayment to Deposit Account No. ___. A Duplicate copy of this sheet is enclosed.

☐ Any additional filing fees required under 37 CFR 1.16.

☐ Any patent application processing fees under 37 CFR 1.17

☐ The Commissioner is hereby authorized to charge payment of the following fees during the pendency of this application or credit any overpayment to Deposit Account No. ___. A duplicate copy of this sheet is enclosed.

☐ Any filing fees under 37 CFR 1.16 for presentation of extra claims.

☐ Any patent application processing fees under 37 CFR 1.17.

☐ The issue fee set in 37 CFR 1.18 at or before mailing of the Notice of Allowance, pursuant to 37 CFR 1.311(b).

(M) ___2/1/92___
date

signature ___John William Inventor___

Form PTO-FB-A510 (10-85)
(also form PTO-1082)

Patent and Trademark Office - U.S. DEPARTMENT of COMMERCE

CHAPTER 7

Step 3: Responding to Office Actions

A year or so after filing an application, you will receive a reply from the Patent Office—the first office action. If the examiner approved of everything, the action will inform you that a patent will be issued upon payment of the issue fee. Normally, however, the action states that various things are wrong and that you must provide a written response which either fixes the problems or explains why the problems do *not* need to be rectified.

THE GENERAL APPROACH

The best way to understand the process of receiving office actions and responding to them is to review a complete history of correspondence between an applicant and the Patent Office concerning a patent application. Such a set of documents is provided in the figures described below.

Figure 7-1 is a sample patent application. Figure 7-2 is a sample first office action sent in response to the patent application. Page 1 of the response is a standard printed checklist used for all office actions. Page 2 is the "Notice of References Cited." This form lists all prior art patents and articles mentioned by the patent examiner in the office action as a basis for rejection of claims. In this case, the examiner cited as references U.S. Patents 4,179,253 and 3,575,165, the Aronberg and Heale patents, respectively. The examiner includes copies of references cited with an office action.

Pages 3, 4, and 5 of Fig. 7-2 are a typewritten explanation of the basis for the examiner's actions, signed by the examiner. The patent applicant must respond to the problem pointed out in each paragraph of this explanation; otherwise, the patent application will be considered abandoned.

Figure 7-3 is the applicant's response to the office ac-

tion presented in Fig. 7-2. Various parts of Fig. 7-3 have been marked with capital letters and are described in the paragraphs that follow.

The first line of the office action, at letter A, gives the patent office serial number for the application, the filing date, the inventor's name (followed by et. al. if more than one inventor), and the attorney's docket number. The attorney's docket number is a number used in your personal filing system and is not something that matters to the patent office.

The name of the patent examiner is given at B and the unit of the patent office in which he or she works is listed at B. In this case, the examiner is Mr. Mednix of art unit 352. Art unit 352 handles all patent applications concerning surgical gowns.

The date the office action was mailed is given at C. This date is critical because it starts the clock running on the time for response to the office action. If you do not respond within the time limit listed at letter D, additional fees will be required and the application will be in danger of being declared abandoned. If you file your response after the allowed period for response (normally 3 months), you will have to file an additional form as seen in Fig. 7-4 and pay the additional fee for late response. See Appendix C to determine the amount of the late fee. *Warning:* If you wait more than six months to respond to an office action, the application will be deemed abandoned and you must file another patent application from scratch.

The first office action is essentially a letter from the patent examiner informing the applicant that one or more claims have been (1) objected to or (2) rejected. The examiner may also object to the drawings or the oath, among other things.

In the case of Fig. 7-2, the examiner rejected claim 2 and 3 of the application because he did not like the use of

RETAINER FOR SURGEON'S EYE GLASSES

BACKGROUND OF THE INVENTION

This invention relates to a retainer for eye glasses and more particularly to a retainer which detachably connects a surgeon's glasses to a surgical hood.

Heretofore, surgeons wearing eye glasses have encountered difficulties in retaining their glasses in place during long hours of tedious surgery. This is especially true in view of the fact that it is often necessary to remove them during surgery without removing the surgical hood. In particular, when a surgeon wearing glasses performs tedious surgery, such as microsurgery, cardio-vascular surgery, emergency (trauma) surgery and the like, he must be able to keep his glasses from slipping due to perspiration or long moments of searching to locate a particular problem and at the same time he must be able to remove them and substitute therefore frames which carry special magnifying lenses. Heretofore, it has been the usual practice for the surgeon to tape the frame of the glasses to his nose and/or temple to keep them from slipping and then remove the tape, leaving the surgical hood in place, when he changes to a frame for supporting special magnifying lenses. Such taping is not only time consuming but is also very uncomfortable and troublesome to the surgeon. Also there is a chance that the surgeon might drop his glasses on the sterilized field or on the patient's open wound where it will cause infection to the patient and contaminate the sterilized field.

1

Fig. 7-1. Patent application.

SUMMARY OF THE INVENTION

In accordance with our invention, we overcome the above and other difficulties by providing a retainer for glasses which is simple of construction and economical of manufacture. Our improved retainer retains a surgeon's glasses positively in place and at the same time permits quick and easy removal thereof without disturbing his operation and/or his surgical hood. Our retainer embodies a flexible tape-like member of a length for one end portion thereof to extend around the nose bridge of a surgeon's glasses and then overlap and be detachably connected to an adjacent portion of the tape-like member, with the other end portion of the tape-like member extending upwardly over a superjacent portion of a surgical hood. adhesive means detachably secures the other end portion of the tape-like member to the hood.

DESCRIPTION OF THE DRAWINGS

A retainer device embodying features of our invention is illustrated in the accompanying drawing, forming part of this application, in which:

Fig. 1 is a perspective view showing our retainer detachably connecting the nose bridge of a pair of glasses to a superjacent portion of the surgical hood; and

Fig. 2 is an enlarged perspective view showing our retainer removed from the glasses and surgical hood.

DETAILED DESCRIPTION

Referring now to the drawing for a better understanding of our invention, we show a flexible tape-like member 10 in the form

2

of a conventional type pressure sensitive tape having an adhesive on one side thereof covered by a detachable connected backing strip 11. One end portion 12 of the tape-like member 10 is of a length to extend around the nose bridge 13 of a surgeon's glasses 14 and overlap an adjacent portion 16 of the tape-like member 10 as shown in Fig. 1. The end portion 12 and the adjacent portion 16 of the tape-like member 10 have cooperating surfaces facing each other, with one surface carrying a plurality of small hook-like members 17 in position to engage a felt-like material 18 carried by the other facing surface. Preferably, the hook-like members 17 are carried by the end 12 with the felt-like material 18 carried by the adjacent portion 16 as shown in Fig. 2. A conventional type fastening device embodying such hook-like members 17 and felt-like material 18 is sold under the trade name "VELCRO".

The tape-like member 10 is of a length for the other end portion 19 thereof to extend upwardly over a superjacent portion of a surgical hood indicated at 21. As shown in Fig. 2, the backing strip 11 covering the adhesive is at the opposite side of the pressure sensitive tape from the side thereof carrying the hook-like members 17 and the felt-like material 18 whereby the adhesive and backing strip 11 carried by the end portion 19 are at the side thereof adjacent the hood 21. Accordingly, after removal of the backing strip 11, the end portion 19 may be attached to the adjacent surface of the hood 21 by pressing the adhesive carrying side of the end portion 19 into contact with the hood. The

3

backing strip 11 of the tape-like member 10 should be a little longer than the end portion 19 so that it will permit the surgeon and/or assistant to peel off the backing strip 11 quickly. In actual practice, we have found that by detachably connecting the end portion 19 to the top portion 22 of the surgical hood 21 in substantial alignment with the nose bridge of the surgeon's eye glasses is satisfactory in every respect for retaining the glasses in place. However, it will be apparent that the tape-like member 10 may be detachably connected to other portions of the surgical hood 21.

From the foregoing description the operation of our improved retainer for surgeon's eye glasses will be readily understood. First, the end portions 12 and 16 of the tape-like member 10 are wrapped around the nose bridge 13 of a surgeon's glasses 14 in position for the hook-like members 17 to engage the felt-like material 18 as shown in Fig. 1. Next, with the surgical hood 21 and the glasses 14 in place, the backing strip 11 is removed from the tape-like member 10 so that the end portion 19 carrying the adhesive may be attached to a superjacent portion of the surgical hood as shown. When the surgeon desires to remove the glasses 14 or change to a frame carrying special magnifying lenses, he disengages the end portion 12 from the adjacent portion 16 whereby the glasses 14 may be easily removed. The nose bridge of the frame carrying special lenses may then be attached between the end portions 12 and 16 as described above without disturbing the surgical hood.

The hook-like members and felt-like material are conventional

4

materials such as nylon or polyolefins.

The adhesive material is made of conventional type adhesives such as epoxy or other synthetic or natural adhesive materials.

Furthermore the materials might be conventional electroconductive materials so that the retainer also functions as a conventional grounding strap grounding the metal eyeglass frame to a conventional grounded hood.

CLAIMS

What we Claim is:

1. A device comprising:

    a. a cap means conformable to a human head,

    b. a pair of eyeglasses, including a nose bridge,

    c. at least one elongated strip means with two end portions, one end portion fastened to the cap means said strip means encompassing and supporting the nose bridge of the eyeglasses, and

    d. at least one end portion releasingly fastened to the strip means.

2. The device of claim 1 wherein the end portion releasingly fastened comprises "Velcro".

3. A device for detachinably connecting a surgeon's eyeglasses having a nose bridge to a head covering comprising:

    a. a head covering means conformable to a human head;

    b. an elongated flexible tape-like member of length for one end portion to extend around said nose bridge and overlap a portion of said tape-like member with the other end

5

extending upwardly,

c.   said elongated flexible member having adhesive fastening
     means at one end portion, and

d.   hook-like members carried by one portion of said elongated
     flexible member cooperating with and engaging a material
     carried by another portion.

ABSTRACT OF THE DISCLOSURE

A retainer for a surgeon's eyeglasses embodies a tape-like member
with one end portion thereof extending around the nose bridge of
the glasses and overlapping and detachable connected to an
adjacent portion of the tape-like member extends alongside and is
detachably connected to a superjacent portion of a surgical hood.

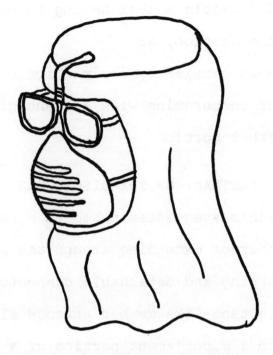

Fig. 1

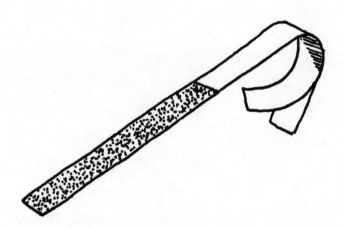

Fig. 2

UNITED STATES DEPARTMENT OF COMMERCE
Patent and Trademark Office
Address : COMMISSIONER OF PATENTS AND TRADEMARKS
Washington, D.C. 20231

SERIAL NUMBER	FILING DATE	FIRST NAMED APPLICANT	ATTORNEY DOCKET NO.
000	8/15/82	Hsu et al	1

Johnny Patentage
Prosperity Ln.
Anywhere, U.S.A. 12121

DOCKETED
~~SEP 2 1983~~

This is a communication from the examiner in charge of your application.
COMMISSIONER OF PATENTS AND TRADEMARKS

RECEIVED
SEP 8 1983

EXAMINER		
Mr. Mednix		
ART UNIT	PAPER NUMBER	
352	2	

DATE MAILED: **09/06/83**

☑ This application has been examined ☐ Responsive to communication filed on _____ ☐ This action is made final.

A shortened statutory period for response to this action is set to expire ~~three~~ month(s), ~~days~~ from the date of this letter.
Failure to respond within the period for response will cause the application to become abandoned.   35 U.S.C. 133

Part I   THE FOLLOWING ATTACHMENT(S) ARE PART OF THIS ACTION:
1. ☑ Notice of References Cited by Examiner, PTO-892.
2. ☐ Notice re Patent Drawing, PTO-948.
3. ☐ Notice of Art Cited by Applicant, PTO-1449
4. ☐ Notice of informal Patent Application, Form PTO-152
5. ☐ Information on How to Effect Drawing Changes, PTO-1474
6. ☐ _____

Part II   SUMMARY OF ACTION

1. ☑ Claims __1 - 3_____ are pending in the application.

     Of the above, claims _____ are withdrawn from consideration.

2. ☐ Claims _____ have been cancelled.

3. ☐ Claims _____ are allowed.

4. ☑ Claims __1 - 3_____ are rejected.

5. ☐ Claims _____ are objected to.

6. ☐ Claims _____ are subject to restriction or election requirement.

7. ☐ This application has been filed with informal drawings which are acceptable for examination purposes until such time as allowable subject matter is indicated.

8. ☐ Allowable subject matter having been indicated, formal drawings are required in response to this Office action.

9. ☐ The corrected or substitute drawings have been received on_____. These drawings are ☐ acceptable;
    ☐ not acceptable (see explanation).

10. ☐ The ☐ proposed drawing correction and/or the ☐ proposed additional or substitute sheet(s) of drawings, filed on _____.
     has (have) been ☐ approved by the examiner. ☐ disapproved by the examiner (see explanation).

11. ☐ The proposed drawing correction, filed_____, has been ☐ approved. ☐ disapproved (see explanation). However,
     the Patent and Trademark Office no longer makes drawing changes. It is now applicant's responsibility to ensure that the drawings are
     corrected. Corrections MUST be effected in accordance with the instructions set forth on the attached letter "INFORMATION ON HOW TO
     EFFECT DRAWING CHANGES", PTO-1474.

12. ☐ Acknowledgment is made of the claim for priority under 35 U.S.C. 119. The certified copy has ☐ been received ☐ not been received

     ☐ been filed in parent application, serial no. _____; filed on _____.

13. ☐ Since this application appears to be in condition for allowance except for formal matters, prosecution as to the merits is closed in
     accordance with the practice under Ex parte Quayle, 1935 C.D. 11; 453 O.G. 213.

14. ☐ Other

PTOL-326 (Rev. 7-82)   EXAMINER'S ACTION

Fig. 7-2. First office action.

TO SEPARATE, HOLD TOP AND BOTTOM EDGES, SNAP–APART AND DISCARD CARBON

FORM PTO-892 (REV. 3-78)	U.S. DEPARTMENT OF COMMERCE PATENT AND TRADEMARK OFFICE	SERIAL NO. 000	GROUP ART UNIT 352	ATTACHMENT TO PAPER NUMBER	2
NOTICE OF REFERENCES CITED		APPLICANT(S) Hsu et al			

U.S. PATENT DOCUMENTS

*		DOCUMENT NO.	DATE	NAME	CLASS	SUB-CLASS	FILING DATE IF APPROPRIATE
	A	4 1 7 9 7 5 3	12/79	Aronberg	2	DIG11	
	B	3 5 7 5 1 6 5	4/71	Heale	128	76000	
	C						
	D						
	E						
	F						
	G						
	H						
	I						
	J						
	K						

FOREIGN PATENT DOCUMENTS

*		DOCUMENT NO.	DATE	COUNTRY	NAME	CLASS	SUB-CLASS	PERTINENT SHTS. DWG	PP. SPEC.
	L								
	M								
	N								
	O								
	P								
	Q								

OTHER REFERENCES (Including Author, Title, Date, Pertinent Pages, Etc.)

R		
S		
T		
U		

EXAMINER Mednix	DATE 12/83	

* A copy of this reference is not being furnished with this office action.
(See Manual of Patent Examining Procedure, section 707.05 (a).)

Serial No. 000

Art Unit 352

1.   Claims 2 and 3 are rejected under 35 U.S.C. 112, second
paragraph, as being indefinite for failing to particularly point out
and distinctly claim the subject matter which the applicant regards
as the invention. Use of trademark "Velcro" makes claim 2 indefinite
since the trademark could stand for different things and its meaning
could change. In claim 3 the means, steps or cooperative relationship
appear to be insufficient to achieve the desired result set forth in
the preamble of the claim.

2.   The following is a quotation of the appropriate paragraphs
of 35 U.S.C. 102 that form the basis for the rejections under this
section made in this Office action:

        A person shall be entitled to a patent unless -

                (b) the invention was patented or

                        described in a printed publication

                        in this or a foreign country or in

                        public use or on sale in this

                        country, more than one year prior to the date

                        of application for patent in the United States.

        Claim 1 is rejected under 35 U.S.C. 102(b) as being clearly
anticipated by Aronberg.

3.   The following is a quotation of 35 U.S.C. 103 which forms the
basis for all obviousness rejections set forth in this Office action:

                A patent may not be obtained though the invention

                is not identically disclosed or described as set forth

                in section 102 of this title, if the differences between

                the subject matter sought to be patented and the prior

art are such that the subject matter as a whole would
have been obvious at the time the invention was made to
a person having ordinary skill in the art to which said
subject matter pertains. Patentability shall not be
negatived by the manner in which the invention was made.

Claims 2 and 3 are rejected under 35 U.S.C. 103 as being
unpatentable over Aronberg in view of Heale.

Aronberg sets forth a head covering with a flexible strip-like
eyeglass support attached to the head covering said strip-like means
fastened to the cap and having one end portion releasingly fastened.

To use a Velcro type fastener or adhesive fastener means
would have been obvious, noting Figs. 7 or 8 of Heale in cooperation
with Aronberg's head covering as merely substitution of one known
and shown fastening means for another. Further it appears obvious that
Velcro, see Fig. 8, could encompass a portion of eye glass rather than
a portion of the cap at 6. Whether the Velcro fastener is facing up
or down is a mere change in direction and not patentable; hence these
claims are held to be ovious to one of ordinary skill in the art.

L.Medniz/pmj
703-557-0000

the trademarked term "Velcro" in a claim. He rejected claim 1 because he thought it claimed something already covered by the Aronberg patent. He also rejected claims 2 and 3 because he thought they covered something implied by the combination of the teachings of the Aronberg and the Heale patent. Details regarding these types of rejections are covered in later sections regarding anticipation, obviousness, and indefiniteness.

Figure 7-3 shows how the patent applicant could respond to the office action of Fig. 7-2. In general, a response must do the following:

1. If the examiner objects to a claim, it means that he does not like the way the claim is written. The response to an objection must usually amend the claim into proper form.

2. If the examiner rejects a claim, it means that he does not think that the invention, as described in the detailed description, is worthy of a claim with that language. In response to a rejection, you must convince the examiner that the invention, as described, really is worthy of that particular claim language, or, you must amend the claim language so that it will be properly supported by what is already in the detailed description section of the application. This often means that you must put more specific language and other limitations into the claims by use of an amendment.

3. In response to objections to the drawings, you must provide any additional drawings required and show the examiner how you propose to fix the problems with the drawings previously submitted. Mark proposed changes in red on a copy of your drawings and return this copy to the examiner with your response. You do not have to file actual, corrected drawings until after you get the notice that the patent has issued.

4. In response to objections and requirements other than the above, various other actions might have to be taken. You can get information regarding these less common problems in advanced texts regarding patent law. See the listing in Appendix D.

Here are two important rules to be followed when responding to office actions:

Rule 1: You must respond in writing, and your response must deal with each point made in the office action. If you fail to do this, your patent application will be considered abandoned.

Rule 2: If you amend or add claims in response to an office action, you must include a written explanation regarding how the amended or additional claims fix the problems pointed out by the examiner.

Before sending in a response to an office action, you should check the response against each paragraph of the office action to assure compliance with rules 1 and 2.

Figure 7-3 shows how a response to an office action should be typed. The format shown is standard and should be used in all cases. The response, like the patent application itself, should be type on 8.5 × 11 inch paper. Double-spaced typing is preferred.

The parts of the standard format response are described below:

- ✍ The first part of the standard format is a page heading with the words "IN THE UNITED STATES PATENT AND TRADEMARK OFFICE."
- ✍ Next comes a block of identifying information. The serial number which must be indicated in this block is assigned to the patent application by the patent office and can be found on page 1 of the office action, as can the information regarding the examiner and art unit.
- ✍ Next is the Amendment section. In this section you request that the patent examiner change the wording of your original patent application.

After addressing the "Honorable Commissioner," you must state why you are filing this paper. In the example, the paper is filed as a response to the office action dated December 9, 1983. It is also possible to file amendments before any office action, for example, if you decide that you want to change or add claims to your application after filing but before getting the first office action. In this case, you should state "This is a preliminary amendment."

CHANGING CLAIM WORDING

After these preliminaries, the real work of amending the claims begins. To show the examiner how the claim is to be changed, you must rewrite the entire claim with words to be deleted enclosed in square brackets and words to be added underlined. Therefore, because the words "for resting a bridge on a human nose" are to be added to claim 1, these words are underlined. On the other hand, where the word "Velcro" is to be deleted, it is written as [Velcro].

When a claim is amended, it is necessary to place either the phrase "(once amended)" or "(amended)" after the number of the claim, provided this is the first time you have amended that particular claim. For further amendments, in response to later office actions, indicate "(twice amended)," "(three times amended)," and so forth.

ADDING CLAIMS

If you desire to add a claim, simply say "Please add the following claim:" and then write the claim to be added. Remember that adding a claim can put you over the three independent/20 total claim allowance limit (as seen in the

<u>IN THE UNITED STATES PATENT AND TRADEMARK OFFICE</u>

Applicant: Joseph W. Applicant   Art Unit: 352

Serial No. 000   Examiner: Mednix

Filed: August 15, 1982

For: Retainer for Surgeon's Eyeglasses

<u>Amendment</u>

Honorable Commissioner of Patents and Trademarks Washington, D.C. 20231

In response to the Office Action mailed December 9, 1983, please amend the claims in the above-identified application as follows:

1. (once amended) A device comprising:

    a. a cap means conformable to a human head,

    b. a pair of eyeglasses, including a nose bridge <u>for</u> <u>resting</u> <u>on a human nose</u>,

    c. at least one elongated strip means with two end portions, one end portion <u>releasably</u> fastened to the cap means, said strip means encompassing and supporting the nose bridge of the eye-glasses <u>and being of sufficient length to support</u> <u>the eyeglasses on the nose when said one end portion is</u> <u>fastened to said cap means,</u> and

    d. at least one end portion releasingly fastened to the strip means.

2. (once amended) The device of claim 1 wherein the end portion releasingly fastened comprises ["Velcro"] <u>cooperating surfaces</u> <u>facing each other with one said surface carrying a plurality</u>

Fig. 7-3. Response to office action.

of <u>small hook like members and the other said surface carrying a felt-like material, said surfaces engageable with each other.</u>

3. (amended) A device for [detachinably] <u>detachably</u> connecting a surgeon's eyeglasses having a nose bridge to a head covering comprising:

    a. a head covering means conformable to a human head;

    b. an elongated flexible tape-like member of length for one end portion to extend around said nose bridge and overlap a portion of said tape-like member with the other end extending upwardly;

    c. said elongated flexible member having adhesive fastening means at one end portion <u>adapted to releasably secure said tape-like member to said head covering means</u>, [and]

    d. hook-like members carried by one portion of said elongated flexible member cooperating with and engaging a material carried by another portion [.] <u>,and</u>

    <u>e.</u> <u>said elongated flexible member being of a sufficient length to allow the eye glasses to be supported on a nose of a wearer when the elongated flexible member encompasses the nose bridge and is attached to said head covering means.</u>

Please add the following claim

4. A device as claimed in claim 3 wherein said elongated flexible

member is made from an electroconductive material whereby the elongated flexible member functions as a grounding strap grounding the eyeglasses to the head covering means.

Remarks

This amendment is in response to the Office Action mailed December 9, 1983. Accompanying this response is a petition for a two month extension of time and the appropriate fee of $75.00 pursuant to 37 CFR 1.17 (b) and 37 CFR 1.136(a). The small entity status has been previously established.

Claims 1-3 have been amended and remain in this application. New claim 4 has been added.

Claims 2 and 3 stand rejected under 35 U.S.C. 112, second paragraph as indefinite. The Examiner held that the use of the trademark "Velcro" in claim 2 made it indefinite since the trademark could stand for different things and its meaning could change. Further, the means, steps or cooperative relationship of claim 3 were held to be insufficient to achieve the desired result set forth in the preamble of the claim.

Claim 2 has been amended to delete reference to the trademark "Velcro". A description of cooperating hook-like members and felt-like material has been substituted for the Trademark "Velcro". Basis for such description can be found in col. 2, lines 2-13 of the specification. Claim 2, as amended, therefore appears to overcome any indefiniteness and withdrawal of the rejection of claim 2 under 35 U.S.C. 112 is requested.

Claim 3 has been amended to include recitation of the flexible tape-like members as adapted to be secured to the head covering means. The rejection of claim 3 as indefinite is seen to be overcome by recitation of this relationship. Withdrawal of the rejection of claim 3 under 35 U.S.C. 112 is requested.

Claim 1 stands rejected under 35 U.S.C. 102(b) as being anticipated by Aronberg.

Claim 1, as amended, recites inter alia, one end portion of the strip means _releasably_ fastened to the cap means and of a sufficient length to support the eyeglasses on the wearer's nose. Neither of these novel features are taught by Aronberg. The strip of Aronberg is _permanently_ attached to the hat by rivets 61,63 and is of a length which does not permit the eyeglass to be supported on the nose. Not only are these newly claimed features not taught by Aronberg, but there is no teaching in any of the prior art of record to suggest providing Aronberg with these features. In view of these reasons, withdrawal of the rejection of claim 1 as anticipated by Aronberg appears in order and is respectfully requested.

Claims 2 and 3 have been rejected under 35 U.S.C. 103 as being unpatentable over Aronberg in view of Heale. The Examiner has held it would be obvious to substitute a Velcro type fastener or adhesive taught by Heale in cooperation with Aronberg's head covering. It was further held that Velcro could encompass a portion of the eyeglasses rather than a portion of the cap at 6 in Heale. Finally, the direction in the Velcro fastener is facing was held as not patentable.

While applicant takes issue with the proposed combination,

even if it were obvious to combine the teachings of Aronberg with Heale as suggested, the claimed features still would not be met. Specifically, there is no teaching in either Aronberg or Heal of providing a flexible strip of a length sufficient to allow the eyeglasses to rest on the nose of a wearer. Heale does not teach a strip used with eye glasses at all and the intent of Aronberg's device is to hold the eye glasses on the cap in the nonuse position. Any attempt to modify the strip of Aronberg to be of a length to allow support of the eye glasses on the nose would not only be the result of hindsight reconstruction of applicant's device, but would also destroy the intended function of the device of Aronberg. A further distinguishing feature of applicant's device is the easy releasability of the end portion of the elongated strip which attaches to the head covering. Inasmuch as there is no reason to make Aronberg's permanent connection the the cap releasable other than from applicant's teachings, it is submitted that it would not be obvious to make the strip of Aronberg easily, releasably fastened to the cap. Regarding any possible combination with Heale, while the substitution of one releasable fastener for another or one permanent fastener for another may be obvious, to substitute an easily releasable type for a permanent type would be contrary to the standard of obviousness without specific support in the prior art for doing so. In view of the above reasons, it is submitted that claims 2 and 3 are not obvious over the combined teachings of Aronberg and Heale and withdrawal of the rejection of these claims is requested.

Newly added claim 4 further defines the material of the flexible strip. Claim 4 depends on claim 3 and it is believed to be allowable for the reasons argued above with respect to claim 3. Furthermore, none of the prior art of record even remotely suggests the electroconductive material claimed.

In view of the above reasons, it is submitted that claims 1-4 are allowable and applicant respectfully requests an early notice to such effect.

Respectfully submitted,

Joseph W. Applicant

schedule of patent fees in Appendix C), so you might have to send in extra money before the patent office will let you add the claim.

DELETING CLAIMS

If you desire to eliminate a claim, simply say "Kindly strike claim 3." If you strike a claim for each claim you add, no extra fees need be paid.

AMENDING THE SPECIFICATION

Sometimes it is necessary to amend the Summary of the Invention or the Detailed Description portions of the application. These portions, taken together, are referred to as the *Specification* portion of the application. To amend something here, you must state that you are making a change in the Specification, point to the exact page and line in the patent application, and tell the examiner exactly what change is to be made.

Examples:

"In the Specification at page 2, line 14: strike the misspelled word—*votage*—and replace it with the word "*voltage*."

"In the Specification at page 5, line 24: after the word —*light*—, add the word "*bulb*."

You may make routine grammatical or spelling changes in the specification, but you may not amend in "new matter" that provides further explanation of your invention. This is not permitted because it would allow new material to get the benefit of an earlier filing date.

The Remarks section of the response (letter D) follows the Amendments section. Although not every response has to have an Amendments section, you must always include remarks. Even if you do not change the content of an application, you must at least explain to the patent examiner why you think his rejections or objections are unjustified.

The first paragraph of the Remarks section should state the reason you are sending this particular paper to the patent office. For example: "This paper is sent in response to the office action dated . . . " When adding claims for which an additional fee is required, state that you have paid the fee and have previously filed an application for small entity status; hence you are qualified to pay according to the lower fee schedule.

The second paragraph should state what claims are active in the application. Active claims are those that have not been stricken or cancelled. Thus, even if a claim has been rejected or objected to, it is still active. You should also state what claims have been added, stricken, or amended by the Amendments portion of the response.

Next, paragraphs should be included explaining the importance of each added claim or amendment to a claim. You should explain, for each claim, why the added or different language makes the claim more deserving of allowance than

previously rejected or objected to claims. Explain how the changed language overcomes particular objections raised by the examiner. Since all language in each claim must be supported by similar language in the Specification, that is, in the Summary or Detailed Description section, include a sentence pointing out where language new to the claims can be found in the Specification.

After explaining additions and amendments, discuss each ground of rejection or objection mentioned in the office action. Begin each discussion by restating what the examiner did and summarizing the reasons he gave for doing it. Then present your side of the story. If you have amended a claim, explain in detail why the claim, as amended, no longer has the same shortcomings as it had previously. If you have presented an affidavit as evidence in support of a claim, for example, an affidavit proving an invention date prior to that of a reference patent, discuss this together with your other arguments. You must do more than simply include the affidavit with the response.

You can get ideas for arguments from the discussion of the various grounds for rejection and objection presented in the next section.

The last paragraph of the response should request that the examiner reconsider his rejections or other actions and allow the outstanding claims.

Finally, you must personally sign the response. If an invention is being applied for by more than one inventor without an attorney, *all* must sign the response. Observe how Fig. 5-3 complies with the foregoing suggestions.

RESPONDING TO LATER OFFICE ACTIONS

The foregoing discussion concerned a response to a first office action. After you respond to the first office action, the examiner will reconsider your application in light of the response and decide whether to issue a patent. If he feels that there is still something wrong, he will send a second office action to which you must also respond.

If the second office action cites different prior art or presents some new grounds for rejection of your claims, you may respond by amending the claims. When responding to a non-final second office action, although you may not radically change the nature of the claims with an amendment, you may add additional limitations or rewrite them in order to overcome the specific art or objections made in the office action.

If the second office action does not contain any new references or grounds for objection or rejection, the examiner will declare his action to be "final." This means that you no longer have a right to amend the claims. At this point, you must either appeal to a higher authority, that is, the Patent Office Board of Appeals, for an order reversing the examiner's decision, or, file a continuing patent application which more or less lets you start from scratch. Your response to a final office action must be either (1) a written notice of appeal and a written argument, called a *brief*, explaining the basis for the appeal, or (2) the filing a second patent application, known as a *continuation*. Both of these options require additional fees.

Appeals

The procedure for appeals is beyond the scope of this book. It is recommended that you consult volume 5 of Kayton, "Patent Preparation and Prosecution Practice," for detailed information. This book is generally available at the law school libraries of major universities.

Continuing Patent Applications

A continuing patent application allows you to refile for a patent and begin the patent application again. The patent office gets another filing fee, and you get a fresh chance to amend your claims and present written arguments to the patent examiner. One procedure for filing a continuing patent application is exactly the same as that for filing the original patent application. However, on the first page of the application, you should include a new section entitled "References to Related Applications." This section should contain the following statement: "This application is a continuation of the application with serial number _____ which was filed on _____. The earlier filing date of this application is hereby claimed."

At the time you file a continuing application, you can also file an amendment entitled "preliminary amendment," which adds new claims or changes the claims carried over from the previous application. Because the major reason for filing a continuing application is to give yourself another chance to amend the claims and present arguments to the patent examiner, it makes sense to file such a preliminary amendment with any continuation. Otherwise, you merely get the same rejection from the same patent examiner and the second application will get you no farther than the first.

From the foregoing discussion, you should learn this lesson: respond carefully and thoroughly to the first office action. This response might be your only opportunity to amend or add claims into an acceptable form prior to getting a final action. By taking the time to prepare a well thought out response to the first action, you can avoid the need for filing an appeal or continuation later.

DEALING WITH REJECTIONS, OBJECTIONS AND REQUIREMENTS

There are two types of rejection: (1) rejections based on prior art and (2) non-art rejections, based on legal technicalities. These are known as *technical rejections*. This section explains how to respond to each type of rejection.

Rejections Based Upon Prior Art (Section 102 and 103 Rejections)

Prior art rejections are based on Sections 102 or 103 of the Patent Statute. Because these sections appear in volume 35 of The United States Code, these sections, referred to as 35 U.S.C. 102 and 35 U.S.C. 103, are reproduced in Appendix B.

Sections 102 and 103 list various situations in which the grant of a patent is not authorized. The more common of these are discussed below:

SECTION 102(a) REJECTIONS

Section 102(a) is probably the most common prior art rejection. Section 102(a) prohibits a patent if the invention was known or used by others before the applicant invented it. Mere knowledge or use of the invention in the U.S. is sufficient. Outside the U.S., mere knowledge or use is not adequate: the patent examiner must refer to a patent or printed publication.

The patent examiner rejects a claim under 102(a) by including a separate paragraph in the office action. In this paragraph the examiner quotes Section 102 and gives a citation to a prior art publication or patent. Or, if the rejection is based on prior knowledge or use, evidence is cited of such knowledge or use as of a certain date. The date of the printed publication, patent knowledge, or use is known as the *effective date of the reference*.

There are two ways to overcome a Section 102(a) rejection.

> The inventor can prove that he made his invention before the effective date of the reference; or,
> The inventor can argue that the reference does not show the same invention that he is claiming in the patent application.

How to Prove Invention Before the Effective Date of the Reference. The key date in Section 102(a) is the date of invention, not the date the patent application was filed. In theory, the date of invention is the date the inventor first formed a fully complete conception of the invention in his mind. In practice, the date of invention is the earliest date the inventor can prove that he had the idea—by means of a laboratory notebook signed by witnesses, for example.

A patent application, as filed, contains nothing that tells the examiner about the actual date of invention. So how does the examiner know if any given reference patent or publication existed before the date the applicant conceived the invention? Answer: The examiner doesn't worry about the actual date of invention. If the reference is dated before the filing date of the patent application, a 102(a) rejection is automatically issued.

This seems unfair! Under Section 102, the inventor is entitled to credit for the date he conceived the invention, not the filing date. If the reference patent or publication is dated after the date of invention, the inventor should not be penalized, even if that patent application was filed later. What can the inventor do to ensure that proper credit is given for the earlier invention date?

The answer is to "swear back" of the date of the reference. This means that the inventor signs an affidavit swearing to his actual date of invention and sends in this affidavit along with his response to the office action. A sample affidavit is presented in Fig. 7-5.

An affidavit claiming an earlier invention date, also known as a Rule 131 affidavit, must do more than simply assert the date of invention. The affidavit must set forth facts that demonstrate the earlier date of invention. In order to prove facts which they might someday need to be put into a Rule 131 affidavit, serious inventors keep a bound laboratory notebook in which they write down all the important ideas that occur to them. Each time an important idea is entered into the book, a witness who is able to understand the idea is asked to sign and date the page. Copies of pages from this notebook accompany future affidavits in order to back up invention dates claimed in the patent affidavit. If an affidavit demonstrates a date of invention earlier than the date of a prior art reference, the patent examiner will drop a rejection based on the reference.

For the purposes of rule 131, an invention is considered to have two parts: (1) the conception and (2) the reduction to practice. *Conception* is the formation of a complete idea of how the invention is to work in the mind of the inventor. *Reduction to practice* is actual building and testing of the invention. Alternatively, reduction to practice can be accomplished "constructively," that is, by the filing of a patent application, prior to any actual building and testing.

It is not enough for the 131 affidavit to show facts that merely prove that the inventor had an idea on such-and-such a date. The law rewards an early conception date only if conception is followed up by a diligent effort to get the invention either built or filed in a patent application. Thus, to prove an earlier date of invention, a rule 131 affidavit must prove (1) conception of the invention and (2) reasonable diligence by the inventor in reducing the conception to practice. If conception prior to the reference date is shown to be backed up with diligent effort leading to a reduction to practice, actual or constructive, the reference is overcome.

If the invention was actually reduced to practice before the patent application was filed, the rule 131 affidavit should describe when and where the invention was built and how it was tested. The affidavit should also set forth facts showing that the inventor worked diligently from a date before the date of the prior art reference until the time of reduction to practice. Records of time spent in a laboratory, receipts for materials and supplies, witnessed lab notebook entries and similar items should be described in the affidavit and attached as photocopy exhibits. Every week should be accounted for.

If the invention was conceived but never built prior to filing for a patent, then *diligence* must be shown pertaining to the constructive reduction to practice—that is, the patent filing. The affidavit must set forth facts proving the conception date and, additionally, facts showing that the inventor took prompt, diligent steps to get a patent application filed.

What is *diligence*? There is no exact definition. However, it generally means that the inventor engaged in a continuous, week-by-week effort to get his invention reduced to practice; or, if there was some time gap in his efforts, that this time gap was for a good reason. Thus, if the inventor took two months vacation in Florida to "get away from it all" during which time he did not work on the invention, then he was not diligent. However, if he suffered a heart attack and was under medical orders not to work for two months, his lack of diligence during this time will be excused—provided that the 131 affidavit sets forth the facts concerning the illness.

The Other Way to Overcome References. If you cannot show an invention date earlier than the date of the reference, you must use the second way to overcome a 102(a) rejection: convince the examiner that the reference does not cover the same invention that is covered by the rejected claim. There are two ways that you can do this:

✍ Put a written argument into your response that explains the difference or shortcomings of the reference; or
✍ Amend the rejected claim to put in some language that clearly distinguishes the claimed invention from the reference.

NON-ENABLING OR INOPERABLE REFERENCES. The most common type of argument concerning a reference has to do with whether it shows the same invention as you are claiming or something else. Before a reference can be deemed to anticipate your invention, it must at least (1) accomplish the same result (2) using the same method or mode of operation. When making arguments concerning the reference, you should consider whether your invention (1) is more economical to build or operate; (2) is useful for some purpose not achievable with the reference invention; (3) works substantially better; or (4) avoids some undesirable side effect of the previous invention. In general, any difference in useful properties or construction that is new and non-obvious can be used to distinguish the reference.

If your invention really does have some special advantage not present in the reference, it is probably true that it is built in a different way or has some other physical feature that is different from the reference invention. If so, there should be language in your claims that defines this physical difference. You should consider amending the claims so that this language is present, if necessary. Discuss this language thoroughly in the remarks section of your response and explain why the reference does not meet this specific language of your claims.

One way a reference can be different from your invention is that it does not really teach how to make and use what it shows. To be valid basis for rejecting a claim, a reference must put an invention "in possession of the public." This means that a person of ordinary skill in the field of technology, upon reading the reference, must be able to build the

IN THE UNITED STATES PATENT AND TRADEMARK OFFICE

Applicant: Joseph W. Applicant   Art Unit: 352

Serial No.: 000   Examiner: Mednix

Filed: August 15, 1982

For: Retainer for Surgeon's Eyeglasses

Petition for Extension of Time

Honorable Commissioner of Patents and Trademarks

Washington, D.C. 20231

It is requested that an extension of time to file an amendment to the official Action of December 9, 1983 regarding the above identified application be granted for a period of two (2) months up to and including May 9, 1984.

The fee of $75.00 is enclosed (small entity status has been previously established).

Respectfully submitted,

Joseph W. Applicant

Joseph W. Applicant

Fig. 7-4. Request for extension of time.

thing shown. If the reference does not make this possible, it is considered "inoperable" or "non-enabling," and therefore not valid. Your response must explain what important information the reference fails to teach. If you have performed experiments according to the teachings of the reference with negative results, you must also include a full description of your experiments and details regarding the poor results in a rule 132 affidavit. A rule 132 affidavit has the same general format as a rule 131 affidavit and is signed and submitted in the same way. However, instead of describing facts pertaining to your date of invention, the rule 132 affidavit is a general vehicle for getting other types of facts before the examiner for his consideration.

If you wish, you can simultaneously argue that your present claims are allowable and also add additional claims to the application. This might cost more because a fee is charged for adding claims past a certain number, but it does give you a maximum chance for success because the additional claims might have language that is more acceptable to the examiner.

A COMMON MISTAKE. A common mistake in responding to rejections is to make arguments comparing the reference with the entire patent application, rather than the exact language of the rejected claim. Often the examiner knows that an application contains a patentable invention and that there are great differences between what is in the application and what is taught by the prior art reference. However, he rejects a certain claim anyway. How can the examiner, who knows that a difference is significant still reject a certain claim?

Even though the claim covers the new invention, it is also broad enough to cover something that is old. Because the claim covers old material, it must be rejected.

Example: A patent application is filed concerning a new type of light bulb filament made of dried kudzu plant fibers coated with carbon. This invention is the solution to the energy crisis: it requires only one-tenth the electrical power as previous filaments for the same amount of light output. Claim 1 of the patent application is this:

1. A light bulb comprising: a) plant fibers formed into a filament; and b) carbon coated over said filament in an amount effective for electrical conduction.

This language covers the kudzu filament invention.

In our scenario, when the inventor receives a rejection of claim 1 from the Patent Office, he is furious. The rejection cites Thomas Edison's original light bulb patent as prior art. But Edison's bulb used 10 times as much energy! It is clearly not the same invention. How can this rejection be fair?

The answer is that the language of claim 1 also covered Edison's invention in addition to the new filament because the language of claim 1 was not restricted to kudzu filaments. The term "plant fiber" is also applicable to the cotton fiber used by Edison. Thus, because the language of claim 1 is applicable to what is old, the claim must be rejected.

To overcome the rejection of claim 1, it must be amended to read as follows:

1. (amended) A light bulb comprising: a) kudzu plant fibers formed into a filament; and b) carbon coated over said filament in an amount effective for electrical conduction.

The addition of the word "kudzu" to element (a) now limits the claim such that it does not cover anything invented by Edison. The 102(a) rejection has been overcome.

SECTION 102 REJECTIONS

Section 102(b) bars the grant of a patent if the invention was patented or described in a printed publication in this or a foreign country or in public use or on sale in this country more than one year prior to the date of the application for patent in the United States.

Unlike section 102(a), section 102(b) does not depend on the date of invention. No matter what the invention date, a printed publication or patent issued in the U.S. more than 12 months before a patent application is filed will absolutely prevent the grant of a patent. This is true even if the inventor himself was the author of the reference publication or patent or if the inventor himself was the one to offer the invention for public use or sale.

Under section 102(b), there is no place for "swearing back" of the reference filing date. However, if the publication does not show the exact invention as in the application, you may amend the claims so they do not overlap what is in the reference supporting the 102(b) rejection.

SECTION 102(d) REJECTIONS

Section 102(d) bars the grant of a patent if the invention was first patented or caused to be patented or was the subject of an inventors certificate, by the applicant or his legal representatives or assigns in a foreign country prior to the date of application in this country on an application for patent or inventors certificate filed more than 12 months before the filing of the application in the United States.

If the inventor: (1) files for a patent in a foreign country more than 12 months before the U.S. application is filed; and, (2) if the foreign patent is granted before the U.S. application is filed; then, Section 102(d) bars grant of a patent.

If you file for a U.S. patent before filing abroad, this problem will never arise. Another law, 35 U.S.C. 183 requires that U.S. citizens not file in a foreign country until at least 6 months after the U.S. application is filed. This allows Defense Department officials time to screen all patent applications for ideas useful to the military. The penalty for violation of this is forfeiture of right to a patent.

If your claims are rejected for this reason, and if you really did file abroad too early, there is not much you can do about it by way of amendment or otherwise. The best cure in this case is prevention: file in the U.S. first.

SECTION 102(e) REJECTIONS

Section 102(e) bars the grant of a patent if the invention was described in a patent on an application for patent by another filed in the United States before the invention thereof by the applicant for a patent . . .

This says that a patent will be barred if, before the inventor came up with the idea, it had already been described in a patent application filed by another.

As a practical matter, for purposes of this section, the examiner will assume that the phrase "before the invention thereof" means any date before the date the patent application was filed, unless the inventor "swears back" to an earlier date. This is done by filing a rule 131 affidavit, in the same manner as has been described for 102(b) rejections. As with 102(b) rejections, you must either (1) swear back of the date of the reference or (2) argue that your invention is not the same as that in the reference patent.

SECTION 103 REJECTIONS: THE CONCEPT OF OBVIOUSNESS

Section 102 rejections apply if the invention shown in the reference publication or patent is of an invention more or less identical to that claimed in the patent application. This means that the reference, standing alone, teaches something that is within the territory marked out by the rejected patent claim. This is called the "four corners" doctrine: The invention must be taught within the four corners of a single patent or publication for there to be a valid section 102 rejection.

Conversely, section 103 allows a rejection to be based upon a combination of references—the four corners doctrine does not apply. The examiner may cite two or more references, none of which teach the exact invention but which, when put together, make the invention "obvious." However, the examiner can combine references only if combining the references is reasonable.

If the examiner rejects a claim as obvious in view of a combination of references, look first at the dates of the references. If you can knock out one of the references by "swearing back" to an earlier date, this will dispose of the rejection. Use a rule 131 affidavit as previously described to do this.

Assuming that you can't swear back to an earlier invention date, argue that it is not reasonable to combine the references, or that, if it is reasonable, the combination teaches something other than the claimed invention. A combination of references can be unreasonable for various reasons:

- The references are from unrelated fields of technology.
- None of the references suggests that such a combination is possible or provides any motivation for the combination.
- The references can be combined, but the combination could just as well be made in a way different from that which results in the invention.
- The references can be combined, but only after extensive experimentation and testing. By themselves, the references do not teach exactly how to combine their teachings into the present invention. It is only in hindsight that a way to combine the references can be seen.

- The claimed invention has surprising advantages that are not foreseeable from any of the references.

Here are some examples that illustrate how to respond to section 103 rejections.

Example 1: The patent application claims
1. A light bulb comprising
   a) kudzu plant fibers formed into a filament; and
   b) a coating of carbon, covering said filament, in an amount effective for conducting electricity.

The examiner searches all the patents and publications in his files, but cannot find any that involve kudzu plant fibers. Because there is no reference that teaches a more or less identical invention within its four corners, the examiner cannot issue a rejection under section 102.

However, the examiner finds Edison's light bulb patent that teaches cotton fibers coated with carbon. Furthermore, there is a Japanese patent that discloses a waterproof basket made out of kudzu fibers coated with carbon. Therefore, a section 103 rejection is issued based upon a combination of these two references. How should you respond to this 103 rejection?

The best response would be to argue that it is not reasonable to combine a reference from the art of basket weaving with a reference from the art of light bulb making. People who make light bulbs are not usually skilled in the art of basket making and would not think to look up basket making patents when looking for a way to build a light bulb. Hence, it could not be "obvious" to bulb makers to combine the teachings of such a patent with Edison's light bulb patent. Therefore, the references are not properly combined and the examiner should reconsider his section 103 rejection.

Example 2:
Assume the same facts as in Example 1, except the Japanese patent is for carbon-coated kudzu fibers as a filament in a radio tube. Because the Japanese patent and Edison's patent are from the same field of art—electrical engineering—it will likely be considered reasonable to combine these teachings, and the section 103 rejection is proper. What can you do next?

First, you should argue that vacuum tube filaments are not the same as light bulb filaments. The purpose of the tube filament is to provide heat, whereas the purpose of the light bulb filament is to provide light. Therefore, because these filaments do different things, the ordinary technician wishing to build a light bulb would not think to get his instructions from a vacuum tube patent.

Next, look for a way to change the claim so that it includes a limitation that distinguishes it from the Japanese patent. A good technique is to add an additional claim to the patent that is identical to the rejected claim except for one additional limitation. This way you can argue for allowance of the first claim and, failing that, allowance of the new claim containing an additional limitation. Here is how you might proceed. Upon a careful reading of the detailed description in the Japanese patent, you notice that the Japanese vacuum tube filament is coated with graphite dust whereas the filament in your patent application is coated with charcoal. The difference between graphite and charcoal will probably not be enough to demonstrate a new invention without further proof—the examiner will assume that these two forms of carbon work the same way. Therefore, you must prove to the examiner that this difference in the type of carbon is significant. To do this, build a light bulb with the same filament as in the Japanese radio tube. You are in luck—graphite dust does not work on the light bulb filament, because it gives only ¼ as much light. Now you have the information you need to overcome the obviousness rejection.

To take advantage of this information, you must do three things:

✍ Amend the claim or add a new claim so that charcoal rather than just carbon is specified.

✍ File an affidavit, known as a rule 132 affidavit, which describes your test of graphite filaments and explains that they do not work. As described previously, a rule 132 affidavit can be prepared using the same format as a 131 affidavit. See Fig. 7-5.

✍ Discuss the contents of the rule 132 affidavit in the Remarks section of your response. Explain how the evidence presented in the affidavit overcomes the examiner's rejections.

Example 3: The rejected claim is
1. A battery comprising:
   a) a liquid container;
   b) a magnesium electropositive electrode inside the container and having an exterior terminal;
   c) a fused cuprous chloride electronegative electrode; and
   d) a terminal connecting with said electronegative electrode.

The examiner is unable to find a single prior art battery with these exact electrodes. Therefore, section 102 does not apply. However, the examiner has found a reference to a previous battery with a silver chloride electrode and another reference to a battery with a zinc electrode. Furthermore, he has found references that say cuprous chloride is the equivalent of silver chloride and that magnesium is the equivalent of zinc. The combination of these references, the examiner says, makes obvious the claimed battery. This is the basis for the rejection under section 103. All of the references are dated before the invention in the patent application was made.

How should you respond to this rejection? Because you are unable to swear back of any of the references, you must respond by proving that it is not proper to combine references in the manner the examiner is doing. In the actual patent case from which this example is taken, the patent owner overcame the argument that his invention was obvious with the following arguments:

1. *Unforseen Advantages*. The new battery could be readied for operation by merely adding water, whereas the batteries described in the references could not be. The ability to use water is a special advantage that would not be obvious to persons reading the prior art. Therefore, the claimed battery is non-obvious and the rejection is improper.

2. *Inoperable References*. The inventor attempted to build the battery using methods described in the references. These methods did not work—in fact, an explosion resulted. Because the references do not teach how to make and use the claimed invention, they do not make the claimed battery obvious.

3. *Skepticism of Experts*. Noted experts in the field of battery manufacture have stated that it is impossible to make a battery with these electrodes and with water as the working fluid. Because the experts say it is impossible, the invention is not obvious to the ordinary person of skill in the art of battery making. This makes the battery, by definition, non-obvious.

In order to confirm the arguments 1 through 3 listed above in your response to the examiner's office action, it is necessary to bring extra facts to his attention. These facts are presented to the examiner by writing them into a rule 132 affidavit and attaching this affidavit to your response in the manner previously described.

Other Ways to Show Non-obviousness. Other arguments against obviousness rejections are based upon the notion "If it is so obvious, then why didn't someone do it before?" If something (1) is a great idea, and (2) is obvious, then someone should have done it a long time ago. The fact that this good idea has not been in use proves that the references cited by the examiner do not make it obvious. Here are the arguments:

❢ Great commercial success for the invention soon after it was put on the market.
❢ The fact that the invention solves a problem that others have long complained about; that is, the invention meets a long felt need.
❢ Unsuccessful attempts by others to solve the same problem.
❢ The current use of more expensive solutions to the same problem. (This argument is especially effec-

BEFORE THE UNITED STATES PATENT AND TRADEMARK OFFICE

In re the application of)
Joseph W. Cool, Inventor)
Serial No. 999,999)
Filed: Jan. 3, 1988)   Examiner: I.M. Pickey
FOR: RADON ENHANCED LIGHT BULB)
)
_____)

Affidavit Under Rule 131

Honorable Commissioner of Patents and Trademarks
Washington, D.C. 20231

Commissioner:

   1.   I am the inventor.

   2.   The Examiner has cited as prior art a patent issued June 3, 1987.

   3.   I conceived the idea of a radon-powered light bulb prior to January 2, 1987. On that day I made an entry in my lab notebook that completely described how to build the invention. A copy of the notebook page, signed and witnessed by two of my friends who are engineers, is attached to this affidavit.

   4.   On January 4, I commenced the work of actually building and testing the invention. I worked on the invention at least 20 hours each week until March 15. Attached are witnessed copies of my lab notebook pages for January 10, 17, and 30; for February 4,12,27; and for March 5, and 10. The pages describe my various attempts to perfect the operating model.

   5.   On March 15, 1988, my Air Force Reserve unit was called to active duty and I was sent to combat duty in Iran. This prevented me from doing further work on the invention from March 15 to July 15.

Fig. 7-5. Sample affidavit.

6. On July 15, I was relieved from active duty in the Air Force and returned to my normal activities. On July, I purchased additional radon gas for use in building the invention. A copy of the receipt for the radon from AACO-RAD, INC. is attached. On July 25, a complete light bulb was assembled and tested successfully. A witnessed copy of my lab notebook describing this test is attached.

7. All of my work for the purpose of reducing the invention to practice took place in the United States. More particularly, it took place in my basement laboratory located at 1872 Maple Street, Berkely, California.

I have been warned that willful false statements and the like are punishable by fine or imprisonment or both under 18 U.S.C. 1001 and may jeopardize the validity of any patent issuing in this matter. I hereby declare that all statements made upon my own knowledge are true and that all statements made upon information and belief are believed to be true.

_____   _____
Date   Applicant

tive if the claimed invention is now pushing these more expensive solutions off the market.)

These arguments are based upon facts that are not apparent from the patent application itself. Thus, these facts must be brought before the patent examiner by means of an affidavit signed by persons with actual knowledge of the facts, or, if the facts are stated in printed publications, by supplying the examiner with copies of the publications at the time the response is made. As usual, use a rule 132 affidavit to prove these extra facts.

Section 112 Rejections

Section 112 of the patent act requires that (1) the patent application "distinctly claim the subject matter which the applicant regards as his invention." If the examiner rejects a claim because he does not like the way it is written, he will generally cite section 112 as the basis for his action. Poorly written claims are often rejected under 112 for one of the following reasons:

Improper Alternative Expressions. The claim uses the word "or" when describing an element. For example, it would be bad to include such language as "a bolt or nail" in a claim. The examiner will reject this because the language of each claim must describe a particular invention, not several versions at once. Fix this problem by finding a single, general word that allows you to avoid use of the word "or." In this example, you could use the general term "fastener" in place of the alternative expression "bolt or nail."

Omnibus Claims. The claim uses language such as "I claim a light bulb of that seen in Figure 1." This is an improper "omnibus" claim. You should rewrite the claim to delete this language and to include that which describes each element.

Claims Containing Tradenames. Do not use tradenames in your claims. If your invention is held together with acrylic glue, don't use a term such as "Crazy Glue." The ingredients in Crazy Glue are subject to change at the whim of the glue manufacturer. Therefore, your patent claims will contain language that is not definite. To cure this problem, delete the tradename and insert a generic name or physical description of the tradename ingredient.

Other Rejections

The most common types of rejections have been described. Many other less common rejections are possible. If you run into one of these, consult one of the reference works listed in Appendix D.

CHAPTER 8

Step 4: Responding to the Notice of Allowance

If the examiner is persuaded that a patent should be issued on your application you will receive a document entitled "Notice of Allowance."

In order to get the patent, you must respond to the notice of allowance within three months by (1) paying the issue fee for the patent, and (2) submitting formal drawings to replace the rough drawings you previously supplied.

The amount of the issue fee due will be stated in the notice of allowance. Send a check for this amount together with a transmittal letter containing the title and serial number of the application.

In response to your submission of rough drawings, you should have received a list of bonded draftsmen and a sheet of instructions concerning how to make corrections to the drawings. Preparation of formal drawings can be accomplished either by you, a hired draftsman, or by one of the "bonded draftsman" provided on a list by the Patent Office. If you desire to pay one of these draftsman to make your formal drawings, choose one and he or she will arrange to get your rough drawings from the Patent Office to prepare and submit suitable formal drawings.

If you decide to use a local draftsman or prepare formal drawings yourself, be sure that you comply with Patent Office standards. These standards are set forth in the "Guide for Patent Draftsmen," which is reproduced on the following pages. (Editor's Note: The following pages were obtained from an out-of-print publication and are to be used only as a guide.)

Selected Rules of Practice Relating to Patent Drawings

THE DRAWINGS

35 U. S. C. 113. Drawings. *When the nature of the case admits, the applicant shall furnish a drawing.*

81. DRAWINGS REQUIRED. The applicant for patent is required by statute to furnish a drawing of his invention whenever the nature of the case admits of it; this drawing must be filed with the application. Illustrations' facilitating an understanding of the invention (for example flow sheets in cases of processes, and diagrammatic views) may also be furnished in the same manner as drawings, and may be required by the Office when considered necessary or desirable.

No names or other identification will be permitted within the "sight" of the drawing, and applicants are expected to use the space above and between the hole locations to identify each sheet of drawings. This identification may consist of the attorney's name and docket number or the inventor's name and case number and may include the sheet number and the total number of sheets filed (for example, "sheet 2 of 4").

83. CONTENT OF DRAWING. (a) The drawing must show every feature of the invention specified in the claims. However, conventional features disclosed in the description and claims, where their detailed illustration is not essential for a proper understanding of the invention, should be illustrated in the drawing in the form of a graphical drawing symbol or a labeled representation (e.g. a labeled rectangular box).

(b) When the invention consists of an improvement on an old machine the drawing must when possible exhibit, in one or more views, the improved portion itself, disconnected from the old structure, and also in another view, so much only of the old structure as will suffice to show the connection of the invention therewith.

84. STANDARDS FOR DRAWINGS.

(a) **Paper and ink.** Drawings must be made upon pure white paper of a thickness corresponding to two-ply or three-ply bristolboard. The surface of the paper must be calendered and smooth and of a quality which will permit erasure and correction with India ink. India ink, or its equivalent in quality, must be used for pen drawings to secure perfectly black solid lines. The use of white pigment to cover lines is not acceptable.

(b) **Size of sheet and margins.** The size of a sheet on which a drawing is made must be exactly 8½ by 14 inches (21.6 by 35.6 cm.) One of the shorter sides of the sheet is regarded as its top. The drawing must include a top margin of 2 inches (5.1 cm) and bottom and side margins of one-quarter inch (6.4 mm) from the edges, thereby leaving a "sight" precisely 8 to 11¾ inches (20.3 by 29.8 cm.) Margin border lines are not permitted. All work must be included within the "sight." The sheets may be provided with two ¼-inch-diameter (6.4 mm.) holes having their centerlines spaced eleven-sixteenths inch (17.5 mm.) below the top edge and 2¾ inches (7.0 cm.) apart, said holes being equally spaced from the respective side edges.

(c) **Character of lines.** All drawings must be made with drafting instruments or by a process which will give them satisfactory reproduction characteristics. Every line and letter must be absolutely black and permanent; the weight of all lines and letters must be heavy enough to permit adequate reproduction. This direction applies to all lines however fine, to shading, and to lines representing cut surfaces in sectional views. All lines must be clean, sharp, and solid, and fine or crowded lines should be avoided. Solid black should not be used for sectional or surface shading. Freehand work should be avoided wherever it is possible to do so.

(d) **Hatching and shading.** (1) Hatching should be made by oblique parallel lines, which may be not less than about one-twentieth inch (1.3 mm) apart.

(2) Heavy lines on the shade side of objects should be used except where they tend to thicken the work and obscure reference characters. The light should come from the upper

left-hand corner at an angle of 45°. Surface delineations should be shown by proper shading, which should be open.

(e) Scale. The scale to which a drawing is made ought to be large enough to show the mechanism without crowding when the drawing is reduced in reproduction, and views of portions of the mechanism on a larger scale should be used when necessary to show details clearly; two or more sheets should be used if one does not give sufficient room to accomplish this end, but the number of sheets should not be more than is necessary.

(f) Reference characters. The different views should be consecutively numbered figures. Reference numerals (and letters, but numerals are preferred) must be plain, legible and carefully formed, and not be encircled. They should, if possible, measure at least one-eighth of an inch (3.2 mm) in height so that they may bear reduction to one twenty-fourth of an inch (1.1 mm); and they may be slightly larger when there is sufficient room. They must not be so placed in the close and complex parts of the drawing as to interfere with a thorough comprehension of the same, and therefore should rarely cross or mingle with the lines. When necessarily grouped around a certain part, they should be placed at a little distance, at the closest point where there is available space, and connected by lines with the parts to which they refer. They should not be placed upon hatched or shaded surfaces but when necessary, a blank space may be left in the hatching or shading where the character occurs so that it shall appear perfectly distinct and separate from the work. The same part of an invention appearing in more than one view of the drawing must always be designated by the same character, and the same character must never be used to designate different parts.

(g) Symbols, legends. Graphical drawing symbols and other labeled representations may be used for conventional elements when appropriate, subject to approval by the Office. The elements for which such symbols and labeled representations are used must be adequately identified in the specification. While descriptive matter on drawings is not permitted, suitable legends may be used, or may be required in proper cases, as in diagrammatic views and flow sheets or to show materials or where labeled representations are employed to illustrate conventional elements. Arrows may be required, in proper cases, to show direction of movement. The lettering should be as large as, or larger than, the reference characters.

(i) Views. The drawing must contain as many figures as may be necessary to show the invention; the figures should be consecutively numbered if possible, in the order in which they appear. The figures may be plan, elevation, section, or perspective views, and detail views of portions or elements, on a larger scale if necessary, may also be used. Exploded views, with the separated parts of the same figure embraced by a bracket, to show the relationship or order of assembly of various parts are permissible. When necessary a view of a large machine or device in its entirety may be broken and extended over several sheets if there is no loss in facility of understanding the view (the different parts should be identified by the same figure number but followed by the letters, **a, b, c,** etc., for each part). The plane upon which a sectional view is taken should be indicated on the general view by a broken line, the ends of which should be designated by numerals corresponding to the figure number of the sectional view and have arrows applied to indicate the direction in which the view is taken. A moved position may be shown by a broken line superimposed upon a suitable figure if this can be done without crowding, otherwise a separate figure must be used for this purpose. Modified forms of construction can only be shown in separate figures. Views should not be connected by projection lines nor should center lines be used.

(j) Arrangement of views. All views on the same sheet must stand in the same direction and should if possible, stand so that they can be read with the sheet held in an upright position. If views longer than the width of the sheet are necessary for the clearest illustration of the invention, the sheet may be turned on its side so that the two-inch (5.1 cm) margin is on the righthand side. One figure must not be placed upon another or within the outline of another.

(k) **Figure for Official Gazette.** The drawing should, as far as possible, be so planned that one of the views will be suitable for publication in the Official Gazette as the illustration of the invention.

(l) **Extraneous matter.** An inventor's, agent's, or attorney's name, signature, stamp, or address, or another extraneous matter, will not be permitted upon the face of a drawing, within or without the margin, except that identifying indicia (attorney's docket number, inventor's name, number of sheets, etc.) should be placed within three-fourths, inch (19.1 mm) of the top edge and between the hole locations defined in paragraph (b) of this rule. Authorized security markings may be placed on the drawings provided they be outside the illustrations and are removed when the material is declassified.

(m) **Transmission of drawings.** Drawings transmitted to the Office should be sent flat, protected by a sheet of heavy binder's board, or may be rolled for transmission in a suitable mailing tube; but must never be folded. If received creased or mutilated, new drawings will be required. (See rule 152 for design drawings, 165 for plant drawings, and 174 for reissue drawings)

85. INFORMAL DRAWINGS. The requirements of rule 84 relating to drawings will be strictly enforced. A drawing not executed in conformity thereto, if suitable for reproduction, may be admitted, but in such case the drawing must be corrected or a new one furnished, as required. The necessary corrections or mounting will be made by the Office upon applicant's request or permission and at his expense. (See rules 21 and 165)

86. DRAFTSMAN TO MAKE DRAWINGS.

(a) Applicants are advised to employ competent draftsmen to make their drawings.

(b) The Office may furnish the drawings at the applicant's expense as promptly as its draftsmen can make them, for applicants who cannot otherwise conveniently procure them. (See rule 21)

88. USE OF OLD DRAWINGS. If the drawings of a new application are to be identical with the drawings of a previous application of the applicant on file in the Office, or with part of such drawings, the old drawings or any sheets thereof may be used if the prior application is, or is about to be, abandoned, or if the sheets to be used are cancelled in the prior application. The new application must be accompanied by a letter requesting the transfer of the drawings, which should be completely identified.

123. Amendments to the drawing.

(a) No change in the drawing may be made except by permission of the Office. Permissible changes in the construction shown in any drawing may be made only by the Office. A sketch in permanent ink showing proposed changes, to become part of the record, must be filed. The paper requesting amendments to the drawing should be separate from other papers.

(b) Substitute drawings will not ordinarily be admitted in any case unless required by the Office.

DESIGN PATENTS

152. DRAWING. The design must be represented by a drawing made in conformity with the rules laid down for drawings of mechanical inventions and must contain a sufficient number of views to constitute a complete disclosure of the appearance of the article. Appropriate surface shading must be used to show the character or contour of the surfaces represented.

SYMBOLS FOR DRAFTSMEN

Rule 84 (g) states that graphical symbols for conventional elements may be used on the drawing when appropriate, subject to approval by the Office. The symbols and other conventional devices which follow have been and are approved for such use. This collection does not purport to be exhaustive, other standard and commonly used symbols will also be acceptable provided they are clearly understood, are adequately identified in the specification as filed, and do not create confusion with other symbols used in patent drawings.

It should be noted that the American National Standards Institute Inc., 1430 Broadway, New York, N.Y. 10018, publishes a series of publications relating to graphic symbols under its Y32 and Z32 headings, the Office calls attention of patent applicants to these symbols for their consideration and use where appropriate in patent drawings. The listed publications have been reviewed by the Office and the symbols therein are considered to be generally acceptable in patent drawings. Although the Office will not "approve" all of the listed symbols as a group because their use and clarity must be decided on a case-by-case basis, these publications may be used as guides when selecting graphic symbols. Overly specific symbols should be avoided. Symbols with unclear meanings should be labeled for clarification. As noted in Rule 84 (g), the Office will retain final authority to approve the use of any particular symbols in any particular case.

The reviewed publications are as follows:

Y32.2—1970. Graphic Symbols for Electrical and Electronics Diagram $11.50
32.10—Graphic Symbols for Fluid Power Diagrams 3.00
Y32.11—1961. Graphic Symbols for Process Flow Diagrams in the Petroleum and Chemical Industries 2.00
Y32.14—1962. Graphic Symbols for Logic Diagrams 4.75
Z32.2.3—1949 (R1953). Graphic Symbols for Pipe Fittings, Valves and Piping . 2.00
Z32.2.4—1949 (R1953). Graphic Symbols for Heating, Ventilating and Air Conditioning 2.00
Z32.2.6—1950. Graphic Symbols for Heat-Power Apparatus 2.00

NOTES: In general, in lieu of a symbol, a conventional element, combination or circuit may be shown by an appropriately labeled rectangle, square, or circle; abbreviations should not be used unless their meaning is evident and not confusing with the abbreviations used in the suggested symbols. In the electrical symbols an arrow through an element indicates variability thereof, see for example symbols 2, 6, 12; dotted line connection of arrows indicates ganging thereof, see symbol 6; inherent property (as resistance) may be indicated by showing symbol (for resistor) in dotted lines.

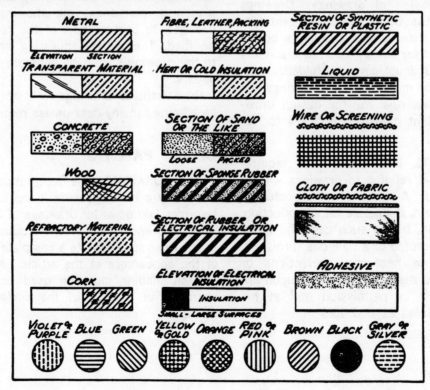

Electrical Symbols

RESISTOR	VARIABLE RESISTOR	POTENTIOMETER	RHEOSTATS	CONDENSERS	GANGED VARIABLE CONDENSERS
1	2	3	4	5	6
INDUCTORS	INDUCTOR ADJUSTABLE CORE	INDUCTOR OR REACTOR POWDERED MAGNETIC CORE	TRANSFORMER SATURABLE CORE	TRANSFORMER AIR CORE	VARIABLE TRANSFORMER
7	8	9	10	11	12
TRANSFORMER MAGNETIC CORE	AUTO-TRANSFORMER ADJUSTABLE	CROSSED AND JOINED WIRES	MAIN CIRCUITS / SHUNT OR CONTROL CIRCUITS	FUSE	COAXIAL CABLES
13	14	15	16	17	18
SHIELDING	BATTERY	THERMOELEMENT	BELL	AMMETER	MILLIAMMETER
19	20	21	22	23	24
VOLTMETER	GALVANOMETER	WATTMETER	SWITCH	DOUBLE POLE SWITCH	DOUBLE POLE DOUBLE THROW SWITCH
25	26	27	28	29	30
PUSH BUTTON TWO POINT MAKE	SELECTOR OR CONNECTOR OR FINDER SWITCH	CIRCUIT BREAKER OVERLOAD	RELAY	POLARIZED RELAY	DIFFERENTIAL RELAY
31	32	33	34	35	36
ANNUNCIATORS SIDE FRONT	DROP ANNUNCIATOR	DRUM TYPE SWITCH OR CONTROL	COMMUTATOR MOTOR OR GENERATOR	REPULSION MOTOR	INDUCTION MOTOR THREE PHASE SQUIRREL CAGE
37	38	39	40	41	42
INDUCTION MOTOR PHASE WOUND SECONDARY	SYNCHRONOUS MOTOR OR GEN. THREE PHASE	MOTOR GENERATOR	ROTARY CONVERTER THREE PHASE	FREQUENCY CHANGER THREE PHASE	TROLLEYS
43	44	45	46	47	48
THIRD RAIL SHOE	RECEIVERS	TRANSMITTER OR MICROPHONE	TELEPHONE HOOK	TELEGRAPH KEY	SWITCH BOARD PLUG AND JACK
49	50	51	52	53	54

Electrical Symbols – continued

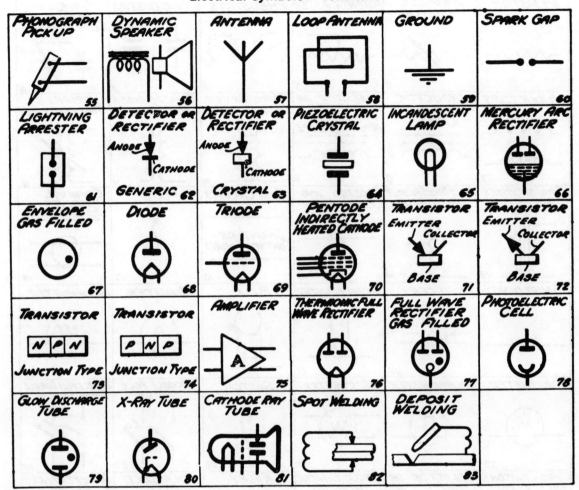

PHONOGRAPH PICK UP 55	DYNAMIC SPEAKER 56	ANTENNA 57	LOOP ANTENNA 58	GROUND 59	SPARK GAP 60
LIGHTNING ARRESTER 61	DETECTOR OR RECTIFIER — ANODE — CATHODE GENERIC 62	DETECTOR OR RECTIFIER — ANODE — CATHODE CRYSTAL 63	PIEZOELECTRIC CRYSTAL 64	INCANDESCENT LAMP 65	MERCURY ARC RECTIFIER 66
ENVELOPE GAS FILLED 67	DIODE 68	TRIODE 69	PENTODE INDIRECTLY HEATED CATHODE 70	TRANSISTOR EMITTER COLLECTOR BASE 71	TRANSISTOR EMITTER COLLECTOR BASE 72
TRANSISTOR NPN JUNCTION TYPE 73	TRANSISTOR PNP JUNCTION TYPE 74	AMPLIFIER 75	THERMIONIC FULL WAVE RECTIFIER 76	FULL WAVE RECTIFIER GAS FILLED 77	PHOTOELECTRIC CELL 78
GLOW DISCHARGE TUBE 79	X-RAY TUBE 80	CATHODE RAY TUBE 81	SPOT WELDING 82	DEPOSIT WELDING 83	

Mechanical Symbols

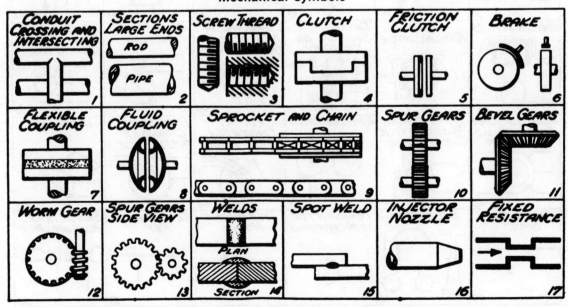

CONDUIT CROSSING AND INTERSECTING 1	SECTIONS LARGE ENDS ROD PIPE 2	SCREW THREAD 3	CLUTCH 4	FRICTION CLUTCH 5	BRAKE 6
FLEXIBLE COUPLING 7	FLUID COUPLING 8	SPROCKET AND CHAIN 9	SPUR GEARS 10	BEVEL GEARS 11	
WORM GEAR 12	SPUR GEARS SIDE VIEW 13	WELDS PLAN SECTION 14	SPOT WELD 15	INJECTOR NOZZLE 16	FIXED RESISTANCE 17

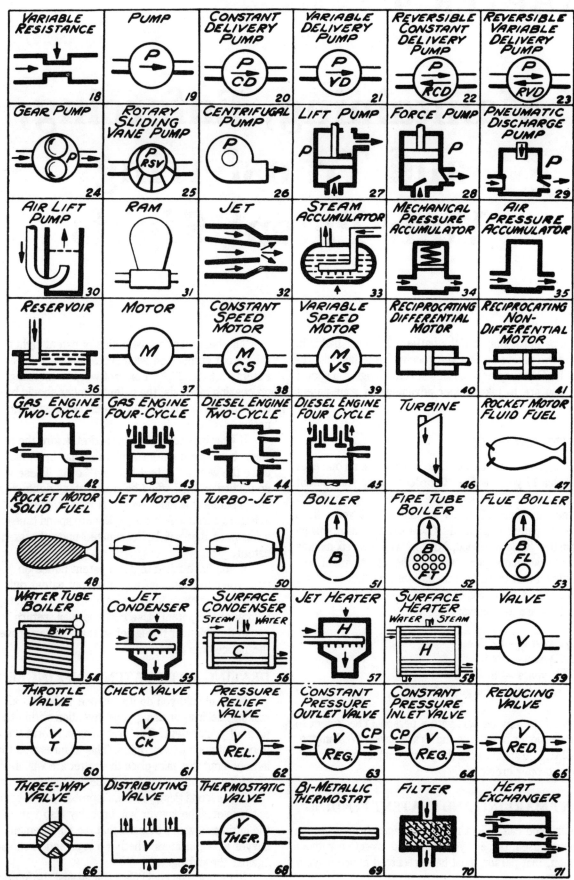

VARIABLE RESISTANCE 18	PUMP 19	CONSTANT DELIVERY PUMP 20	VARIABLE DELIVERY PUMP 21	REVERSIBLE CONSTANT DELIVERY PUMP 22	REVERSIBLE VARIABLE DELIVERY PUMP 23
GEAR PUMP 24	ROTARY SLIDING VANE PUMP 25	CENTRIFUGAL PUMP 26	LIFT PUMP 27	FORCE PUMP 28	PNEUMATIC DISCHARGE PUMP 29
AIR LIFT PUMP 30	RAM 31	JET 32	STEAM ACCUMULATOR 33	MECHANICAL PRESSURE ACCUMULATOR 34	AIR PRESSURE ACCUMULATOR 35
RESERVOIR 36	MOTOR 37	CONSTANT SPEED MOTOR 38	VARIABLE SPEED MOTOR 39	RECIPROCATING DIFFERENTIAL MOTOR 40	RECIPROCATING NON-DIFFERENTIAL MOTOR 41
GAS ENGINE TWO-CYCLE 42	GAS ENGINE FOUR-CYCLE 43	DIESEL ENGINE TWO-CYCLE 44	DIESEL ENGINE FOUR CYCLE 45	TURBINE 46	ROCKET MOTOR FLUID FUEL 47
ROCKET MOTOR SOLID FUEL 48	JET MOTOR 49	TURBO-JET 50	BOILER 51	FIRE TUBE BOILER 52	FLUE BOILER 53
WATER TUBE BOILER 54	JET CONDENSER 55	SURFACE CONDENSER 56	JET HEATER 57	SURFACE HEATER 58	VALVE 59
THROTTLE VALVE 60	CHECK VALVE 61	PRESSURE RELIEF VALVE 62	CONSTANT PRESSURE OUTLET VALVE 63	CONSTANT PRESSURE INLET VALVE 64	REDUCING VALVE 65
THREE-WAY VALVE 66	DISTRIBUTING VALVE 67	THERMOSTATIC VALVE 68	BI-METALLIC THERMOSTAT 69	FILTER 70	HEAT EXCHANGER 71

[24 FR 10375, Dec. 22, 1959]

CHAPTER 9

Maintaining, Enforcing, and Transferring Your Patent Rights

Getting a patent is only half the battle. The other half is in exploiting the patent so that you receive some financial reward for it. This chapter discusses several things you should know if you are to take proper advantage of your patent rights.

MAINTENANCE FEES

Although a patent has a lifetime of 17 years, no one ever said that this 17 years is free. In order to keep your patent in force, you must pay maintenance fees to the government. These fees are due at 3 1/2, 7 1/2 and 11 1/2 years from the date the patent is granted. If you pay the maintenance fee within the "window period," that is, within the six months prior to the due date, you can avoid the late-payment surcharge. If you wait more than six months after the due date, all of your patent rights will be lost. For small entities, maintenance fees are as follows:

```
3 1/2 years — $225
7 1/2 years — $445
11 1/2 years — $670
late payment surcharge — $55
```

These fees may be paid with a personal check sent to the Commissioner of Patents and Trademarks, Washington, D.C. 20231. You should write the patent number on the check to ensure proper credit.

PATENT MARKING AND "PATENT PENDING"

Before you can collect money damages from a person who makes, uses, or sells your patented invention without permission, that person must have received notice of your patent rights.

The best way to give notice of patent rights is to print the patent number on each article that is sold, as in "U.S. Patent 5,345,222." For software containing patented processing methods, the patent number should be displayed on a startup or main menu screen along with the copyright notice.

If your invention goes to market before the patent is issued, but after a patent application has been filed, you can place the words "Patent Pending" on it. These words have no particular legal effect because patent rights do not begin until a patent is actually issued. However, they do discourage imitators because few people want to spend time developing a business that will be terminated upon the grant of your patent.

If you have failed to put a patent notice on your invention, you must notify infringers of your patent rights in some other way, such as, perhaps, a nasty letter accompanied by a copy of the patent. Use certified mail, return receipt requested, to deliver this letter so that you can prove the date the notice was delivered.

DEALING WITH PATENT INFRINGERS

What should you do if someone violates your rights by making, selling, or using your invention within the 17 year term of your patent? The first thing is inform the infringing person of your patent rights. This assures you that acts of infringement that take place thereafter are willful. Willful acts of infringement are subject to heavy money penalties.

The second thing you should probably do is contact a patent lawyer. The lawyer will probably also write a letter to the infringers with a demand that the infringement stop. The lawyer also considers whether a legal action should be filed against the infringers.

In theory, you could file a suit against the infringers without using a patent lawyer. However, the law requires that all suits for infringement must be filed in a federal district court, not a state court. A federal district court bears no resemblance to one run by Judge Woppner. It's the "big time" with complex procedures and regulations. Bringing a suit for patent infringement into such a court is more difficult than getting a patent in the first place. How to do it is beyond the scope of this book.

If your patent infringement suit looks like a winner, and if the people you want to sue appear to have money, you might be able to find a patent lawyer willing to work on a contingent fee basis (this means that you will not have to pay $100 an hour in lawyer fees). However, if and when you do get money damages, the lawyer gets a percentage, normally in the range of 30 percent. It is higher for small cases and those that wind up in appellate courts after the initial suit is over. The plaintiff, i.e., you, is normally responsible for paying court fees and costs even with contingent fee arrangements.

LICENSING AND ASSIGNING YOUR PATENT

There are two ways to allow others to use your invention in exchange for money payment. First, there is *licensing*. Licensing a patent is similar to renting out a house or apartment. You keep ownership but let someone else use it for a while. A license is merely a written agreement allowing someone to use your patent in exchange for payment of an agreed amount of money or other consideration.

A license can be either exclusive or non-exclusive. If you grant an exclusive license, the person you license has the exclusive right to make, sell, or use the invention. Even you cannot compete with them. You may not grant additional licenses. On the other hand, a non-exclusive license places none of these limitations on you, the patent owner.

A license may or may not permit the right of the second party to make sub-licenses to a third party. If you grant the right to make sub-licenses to the person you license, this person has the right to grant licenses to others. You should make this a point of negotiation.

Licensing is flexible because you can license for limited periods of time or for particular geographical areas. The method used for calculating payment can be anything you and the other party can agree upon.

The second way to market your patent is by *assigning* it. This is like selling your house instead of renting it out. Like a deed for a house, an assignment is recorded in a government office—in this case, at the patent office.

Tax consequences are typically different if you assign a patent rather than license it. It is always prudent to consult an accountant prior to negotiating an assignment or license agreement.

APPENDIX A:

Forms
Suitable For
Photocopying

DECLARATION FOR PATENT APPLICATION

Docket No. _____

As a below named inventor, I hereby declare that:

My residence, post office address and citizenship are as stated below next to my name.

I believe I am the original, first and sole inventor (if only one name is listed below) or an original, first and joint inventor (if plural names are listed below) of the subject matter which is claimed and for which a patent is sought on the invention entitled _____, the specification of which

(check one) ☐ is attached hereto.
　　　　　　☐ was filed on _____ as
　　　　　　Application Serial No. _____
　　　　　　and was amended on _____ (if applicable).

I hereby state that I have reviewed and understand the contents of the above identified specification, including the claims, as amended by any amendment referred to above.

I acknowledge the duty to disclose information which is material to the examination of this application in accordance with Title 37, Code of Federal Regulations, §1.56(a).

I hereby claim foreign priority benefits under Title 35, United States Code, §119 of any foreign application(s) for patent or inventor's certificate listed below and have also identified below any foreign application for patent or inventor's certificate having a filing date before that of the application on which priority is claimed:

Prior Foreign Application(s)　　　　　　　　　　　　　　　Priority Claimed

(Number)	(Country)	(Day/Month/Year Filed)	Yes	No
(Number)	(Country)	(Day/Month/Year Filed)	Yes	No
(Number)	(Country)	(Day/Month/Year Filed)	Yes	No

I hereby claim the benefit under Title 35, United States Code, §120 of any United States application(s) listed below and, insofar as the subject matter of each of the claims of this application is not disclosed in the prior United States application in the manner provided by the first paragraph of Title 35, United States Code, §112, I acknowledge the duty to disclose material information as defined in Title 37, Code of Federal Regulations, §1.56(a) which occurred between the filing date of the prior application and the national or PCT international filing date of this application:

| (Application Serial No.) | (Filing Date) | (Status—patented, pending, abandoned) |
| (Application Serial No.) | (Filing Date) | (Status—patented, pending, abandoned) |

I hereby appoint the following attorney(s) and/or agent(s) to prosecute this application and to transact all business in the Patent and Trademark Office connected therewith:

Address all telephone calls to _____ at telephone no. _____.
Address all correspondence to _____

I hereby declare that all statements made herein of my own knowledge are true and that all statements made on information and belief are believed to be true; and further that these statements were made with the knowledge that willful false statements and the like so made are punishable by fine or imprisonment, or both, under Section 1001 of Title 18 of the United States Code and that such willful false statements may jeopardize the validity of the application or any patent issued thereon.

Full name of sole or first inventor _____
Inventor's signature _____ Date _____
Residence _____ Citizenship _____
Post Office Address _____

Full name of second joint inventor, if any _____
Second Inventor's signature _____ Date _____
Residence _____ Citizenship _____
Post Office Address _____

(Supply similar information and signature for third and subsequent joint inventors.)

Form PTO-FB-A110 (8-83)

133

OMB No. 0651-0011 (12/31/86)

Applicant or Patentee: _____ Attorney's
Serial or Patent No.: _____ Docket No.: _____
Filed or Issued: _____
For: _____

VERIFIED STATEMENT (DECLARATION) CLAIMING SMALL ENTITY
STATUS (37 CFR 1.9 (f) and 1.27 (b)) — INDEPENDENT INVENTOR

As a below named inventor, I hereby declare that I qualify as an independent inventor as defined in 37 CFR 1.9 (c) for purposes of paying reduced fees under section 41 (a) and (b) of Title 35, United States Code, to the Patent and Trademark
Office with regard to the invention entitled _____
described in

    [ ] the specification filed herewith
    [ ] application serial no. _____ , filed _____ .
    [ ] patent no. _____ , issued _____ .

I have not assigned, granted, conveyed or licensed and am under no obligation under contract or law to assign, grant, convey
or license, any rights in the invention to any person who could not be classified as an independent inventor under 37 CFR
1.9 (c) if that person had made the invention, or to any concern which would not qualify as a small business concern under
37 CFR 1.9 (d) or a nonprofit organization under 37 CFR 1.9 (e).

Each person, concern or organization to which I have assigned, granted, conveyed, or licensed or am under an obligation
under contract or law to assign, grant, convey, or license any rights in the invention is listed below:

    [ ] no such person, concern, or organization
    [ ] persons, concerns or organizations listed below*

*NOTE: Separate verified statements are required from each named person, concern or organiza-
tion having rights to the invention averring to their status as small entities. (37 CFR 1.27)

FULL NAME _____
ADDRESS _____
    [ ] INDIVIDUAL   [ ] SMALL BUSINESS CONCERN   [ ] NONPROFIT ORGANIZATION

FULL NAME _____
ADDRESS _____
    [ ] INDIVIDUAL   [ ] SMALL BUSINESS CONCERN   [ ] NONPROFIT ORGANIZATION

FULL NAME _____
ADDRESS _____
    [ ] INDIVIDUAL   [ ] SMALL BUSINESS CONCERN   [ ] NONPROFIT ORGANIZATION

I acknowledge the duty to file, in this application or patent, notification of any change in status resulting in loss of entitlement to small entity status prior to paying, or at the time of paying, the earliest of the issue fee or any maintenance fee
due after the date on which status as a small entity is no longer appropriate. (37 CFR 1.28 (b))

I hereby declare that all statements made herein of my own knowledge are true and that all statements made on information
and belief are believed to be true; and further that these statements were made with the knowledge that willful false statements
and the like so made are punishable by fine or imprisonment, or both, under section 1001 of Title 18 of the United States
Code, and that such willful false statements may jeopardize the validity of the application, any patent issuing thereon, or
any patent to which this verified statement is directed.

NAME OF INVENTOR	NAME OF INVENTOR	NAME OF INVENTOR
Signature of Inventor	Signature of Inventor	Signature of Inventor
Date	Date	Date

Form PTO-FB-A410 (8-83)

134

PATENT APPLICATION TRANSMITTAL LETTER	ATTORNEY'S DOCKET NO.

TO THE COMMISSIONER OF PATENTS AND TRADEMARKS:

Transmitted herewith for filing is the patent application of _____

_____

for _____

_____

Enclosed are:

☐ _____ sheets of drawing.

☐ an assignment of the invention to _____

_____

☐ a certified copy of a _____ application.

☐ associate power of attorney.

☐ verified statement to establish small entity status under 37 CFR 1.9 and 1.27. ———

CLAIMS AS FILED

FOR.	NO. FILED	NO. EXTRA		SMALL ENTITY			OTHER THAN A SMALL ENTITY	
				RATE	FEE	OR	RATE	FEE
BASIC FEE					$170	OR		$340
TOTAL CLAIMS	-20-	*		x$6=	$	OR	x$12=	$
INDEP CLAIMS	-3-	*		x$17=	$	OR	x$34=	$
MULTIPLE DEPENDENT CLAIM PRESENT				+$55=	$	OR	+$110=	$
				TOTAL	$	OR	TOTAL	$

* If the difference in col. 1 is less than zero, enter "0" in col. 2

☐ Please charge my Deposit Account No. _____ in the amount of $ _____ .

☐ A duplicate copy of this sheet is enclosed.

☐ A check in the amount of $ _____ to cover the filing fee is enclosed.

☐ The Commissioner is hereby authorized to charge payment of the following fees associated with this communication or credit any overpayment to Deposit Account No. _____ . A Duplicate copy of this sheet is enclosed.

    ☐ Any additional filing fees required under 37 CFR 1.16.

    ☐ Any patent application processing fees under 37 CFR 1.17

☐ The Commissioner is hereby authorized to charge payment of the following fees during the pendency of this application or credit any overpayment to Deposit Account No. _____ . A duplicate copy of this sheet is enclosed.

    ☐ Any filing fees under 37 CFR 1.16 for presentation of extra claims.

    ☐ Any patent application processing fees under 37 CFR 1.17.

    ☐ The issue fee set in 37 CFR 1.18 at or before mailing of the Notice of Allowance, pursuant to 37 CFR 1.311(b).

date _____   signature _____

Patent and Trademark Office · U.S. DEPARTMENT of COMMERCE

Form PTO-FB-A510 (10-85)
(also form PTO-1082)

AMENDMENT TRANSMITTAL LETTER	ATTORNEY'S DOCKET NO.

SERIAL NO.	FILING DATE	EXAMINER	GROUP ART UNIT

INVENTION

TO THE COMMISSIONER OF PATENTS AND TRADEMARKS:

Transmitted herewith is an amendment in the above-identified application.

Small entity status of this application under 37 CFR 1.27 has been established by a verified statement previously submitted.

A verified statement to establish small entity status under 37 CFR 1.9 and 1.27 is enclosed.

No additional fee is required.

The fee has been calculated as shown below:

	(1) CLAIMS REMAINING AFTER AMENDMENT		(2) HIGHEST NO PREVIOUSLY PAID FOR	(3) PRESENT EXTRA	SMALL ENTITY		OR	OTHER THAN A SMALL ENTITY	
					RATE	ADDIT FEE		RATE	ADDIT FEE
TOTAL	*	MINUS	**	-	x $6=	$		x $12=	$
INDEP	*	MINUS	***	-	x $17=	$		x $34=	$
FIRST PRESENTATION OF MULTIPLE DEP CLAIM					+$55=	$		+$110=	$
					TOTAL ADDIT. FEE	$	OR	TOTAL	$

* If the entry in Col. 1 is less than the entry in Col. 2, write "0" in Col. 3.

** If the "Highest No Previously Paid For" IN THIS SPACE is less than 20, enter "20"

*** If the "Highest No Previously Paid For" IN THIS SPACE is less than 3, enter "3"

The "Highest No Previously Paid For" (Total or Indep.) is the highest number found in the appropriate box in Col. 1

Please charge my Deposit Account No. _____ in the amount of $ _____ . A duplicate copy of this sheet is enclosed.

A check in the amount of $ _____ to cover the filing fee is enclosed.

The Commissioner is hereby authorized to charge payment of the following fees associated with this communication or credit any overpayment to Deposit Account No. _____ . A Duplicate copy of this sheet is enclosed.

Any additional filing fees required under 37 CFR 1.16.

Any patent application processing fees under 37 CFR 1.17

_____
(date)

_____
(signature)

Form PTO-FB-A520 (10-85)
(also form PTO-1083)

Patent and Trademark Office - U.S. DEPARTMENT of COMMERCE

APPENDIX B:

Selected
Patent Statutes

PART II—PATENTABILITY OF INVENTIONS AND GRANT OF PATENTS

CHAPTER 10—PATENTABILITY OF INVENTIONS

SEC.
100. Definitions.
101. Inventions patentable.
102. Conditions for patentability; novelty and loss of right to patent.
103. Conditions for patentability; non-obvious subject matter.
104. Invention made abroad.

§ 100. Definitions

When used in this title unless the context otherwise indicates—
(a) The term "invention" means invention or discovery.

or for the manufacture, use or sale of which substantial preparation was made after the six-month grace period but before the acceptance of a maintenance fee under this subsection, and it may also provide for the continued practice of any process, practiced, or for the practice of which substantial preparation was made, after the six-month grace period but prior to the acceptance of a maintenance fee under this subsection, to the extent and under such terms as the court deems equitable for the protection of investments made or business commenced after the six-month grace period but before the acceptance of a maintenance fee under the subsection.

(d) The Commissioner will establish fees for all other processing, services, or materials related to patents not specified above to recover the estimated average cost to the Office of such processing, services, or materials. The yearly fee for providing a library specified in section 13 of this title with uncertified printed copies of the specifications and drawings for all patents issued in that year will be $50.

(e) The Commissioner may waive the payment of any fee for any service or material related to patents in connection with an occasional or incidental request made by a department or agency of the Government, or any officer thereof. The Commissioner may provide any applicant issued a notice under section 132 of this title with a copy of the specifications and drawings for all patents referred to in that notice without charge.

(f) The fees established in subsections (a) and (b) of this section may be adjusted by the Commissioner on October 1, 1985, and every third year thereafter, to reflect any fluctuations occurring during the previous three years in the Consumer Price Index, as determined by the Secretary of Labor. Changes of less than 1 per centum may be ignored.

(g) No fee established by the Commissioner under this section will take effect prior to sixty days following notice in the Federal Register. (As amended December 12, 1980, Public Law 96–517, sec. 2, 94 Stat. 3017; August 27, 1982, Public Law 97–247, sec. 3, 96 Stat. 317.)

§ 42.   Patent and Trademark Office funding

(a) All fees for services performed by or materials furnished by the Patent and Trademark Office will be payable to the Commissioner.

(b) All fees paid to the Commissioner and all appropriations for

(b) The term "process" means process, art or method, and includes a new use of a known process, machine, manufacture, composition of matter, or material.

(c) The terms "United States" and "this country" mean the United States of America, its territories and possessions.

(d) The word "patentee" includes not only the patentee to whom the patent was issued but also the successors in title to the patentee.

§ 101. Inventions patentable

Whoever invents or discovers any new and useful process, machine, manufacture, or composition of matter, or any new and useful improvement thereof, may obtain a patent therefor, subject to the conditions and requirements of this title.

§ 102. Conditions for patentability; novelty and loss of right to patent

A person shall be entitled to a patent unless—.

(a) the invention was known or used by others in this country, or patented or described in a printed publication in this or a foreign country, before the invention thereof by the applicant for patent, or

(b) the invention was patented or described in a printed publication in this or a foreign country or in public use or on sale in this country, more than one year prior to the date of the application for patent in the United States, or

(c) he has abandoned the invention, or

(d) the invention was first patented or caused to be patented, or was the subject of an inventor's certificate, by the applicant or his legal representatives or assigns in a foreign country prior to the date of the application for patent in this country on an application for patent or inventor's certificate filed more than twelve months before the filing of the application in the United States, or

(e) the invention was described in a patent granted on an application for patent by another filed in the United States before the invention thereof by the applicant for patent, or on an international application by another who has fulfilled the requirements of paragraphs (1), (2), and (4) of section 371(c) of this title before the invention thereof by the applicant for patent, or

(f) he did not himself invent the subject matter sought to be patented, or

(g) before the applicant's invention thereof the invention was made in this country by another who had not abandoned, suppressed, or concealed it. In determining priority of invention there shall be considered not only the respective dates of conception and reduction to practice of the invention, but also the reasonable diligence of one who was first to conceive and last to reduce to practice, from a time prior to conception by the other (Amended July 28, 1972, Public Law 92–358, sec. 2, 86 Stat. 501; November 14, 1975, Public Law 94–131, sec. 5, 89 Stat. 691.)

§ 103. Conditions for patentability; non-obvious subject matter

A patent may not be obtained though the invention is not identically disclosed or described as set forth in section 102 of this title, if the differences between the subject matter sought to be patented and the prior art are such that the subject matter as a whole would have been obvious at the time the invention was made to a person having ordinary skill in the art to which said subject matter pertains. Patentability shall not be negatived by the manner in which the invention was made.

§ 104. Invention made abroad

In proceedings in the Patent Office and in the courts, an applicant for a patent, or a patentee, may not establish a date of invention by reference to knowledge or use thereof, or other activity with respect thereto, in a foreign country, except as provided in sections 119 and 365 of this title. Where an invention was made by a person, civil or military, while domiciled in the United States and serving in a foreign country in connection with operations by or on behalf of the United States, he shall be entitled to the same rights of priority with respect to such inventions as if the same had been made in the United States. (Amended January 2, 1975, Public Law 93–596, sec. 1, 88 Stat. 1949; November 14, 1975, Public Law 94–131, sec. 6, 89 Stat. 691.)

CHAPTER 11—APPLICATION FOR PATENT

§ 111. Application for patent

Application for patent shall be made, or authorized to be made, by the inventor, except as otherwise provided in this title, in writing to the Commissioner. Such application shall include (1) a specification as prescribed by section 112 of this title; (2) a drawing as prescribed by section 113 of this title; and (3) an oath by the applicant as prescribed by section 115 of this title. The application must be accompanied by the fee required by law. The fee and oath may be submitted after the specification and any required drawing are submitted, within such period and under such conditions, including the payment of a surcharge, as may be prescribed by the Commissioner. Upon failure to submit the fee and oath within such prescribed period, the application shall be regarded as abandoned, unless it is shown to the satisfaction of the Commissioner that the delay in submitting the fee and oath was unavoidable. The filing date of an application shall be the date on which the specification and any required drawing are received in the Patent and Trademark Office. (Amended August 27, 1982, Public Law 97–247, sec. 5, 96 Stat. 319.)

§ 112. Specification

The specification shall contain a written description of the invention, and of the manner and process of making and using it, in such full, clear, concise, and exact terms as to enable any person skilled in the art to which it pertains, or with which it is most nearly connected, to make and use the same, and shall set forth the best mode contemplated by the inventor of carrying out his invention.

The specification shall conclude with one or more claims particularly pointing out and distinctly claiming the subject matter which the applicant regards as his invention.

A claim may be written in independent or, if the nature of the case admits, in dependent or multiple dependent form.

Subject to the following paragraph, a claim in dependent form shall contain a reference to a claim previously set forth and then specify a further limitation of the subject matter claimed. A claim in dependent form shall be construed to incorporate by reference all the limitations of the claim to which it refers.

A claim in multiple dependent form shall contain a reference, in the alternative only, to more than one claim previously set forth and then specify a further limitation of the subject matter claimed. A multiple dependent claim shall not serve as a basis for any other multiple dependent claim. A multiple dependent claim shall be construed to incorporate by reference all the limitations of the particular claim in relation to which it is being considered.

An element in a claim for a combination may be expressed as a means or step for performing a specified function without the recital of structure, material, or acts in support thereof, and such claim shall be construed to cover the corresponding structure, material, or acts described in the specification and equivalents thereof. (Amended July 24, 1965, Public Law 89–83, sec. 9, 79 Stat. 261; November 14, 1975, Public Law 94–131, sec. 7, 89 Stat. 691.)

§ 113. Drawings

The applicant shall furnish a drawing where necessary for the understanding of the subject matter sought to be patented. When the nature of such subject matter admits of illustration by a drawing and the applicant has not furnished such a drawing, the Commissioner may require its submission within a time period of not less than two months from the sending of a notice thereof. Drawings submitted after the filing date of the application may not be used (i) to overcome any insufficiency of the specification due to lack of an enabling disclosure or otherwise inadequate disclosure therein, or (ii) to supplement the original disclosure thereof for the purpose of interpretation of the scope of any claim. (Amended November 14, 1975, Public Law 94–131, sec. 8, 89 Stat. 691.)

§ 114. Models, specimens

The Commissioner may require the applicant to furnish a model of convenient size to exhibit advantageously the several parts of his invention.

When the invention relates to a composition of matter, the Commissioner may require the applicant to furnish specimens or ingredients for the purpose of inspection or experiment.

§ 115.   Oath of applicant

The applicant shall make oath that he believes himself to be the original and first inventor of the process, machine, manufacture, or composition of matter, or improvement thereof, for which he solicits a patent; and shall state of what country he is a citizen. Such oath may be made before any person within the United States authorized by law to administer oaths, or, when made in a foreign country, before any diplomatic or consular officer of the United States authorized to administer oaths, or before any officer having an official seal and authorized to administer oaths in the foreign country in which the applicant may be, whose authority is proved by certificate of a diplomatic or consular officer of the United States, or apostille of an official designated by a foreign country which, by treaty or convention, accords like effect to apostilles of designated officials in the United States. Such oath is valid if it complies with the laws of the state or country where made. When the application is made as provided in this title by a person other than the inventor, the oath may be so varied in form that it can be made by him. (Amended August 27, 1982, Public Law 97–247, sec. 14(a), 96 Stat. 321.)

§ 116.   Inventors

When an invention is made by two or more persons jointly, they shall apply for patent jointly and each sign the application and make the required oath, except as otherwise provided in this title.

If a joint inventor refuses to join in an application for patent or cannot be found or reached after diligent effort, the application may be made by the other inventor on behalf of himself and the omitted inventor. The Commissioner, on proof of the pertinent facts and after such notice to the omitted inventor as he prescribes, may grant a patent to the inventor making the application, subject to the same rights which the omitted inventor would have had if he had been joined. The omitted inventor may subsequently join in the application.

Whenever through error a person is named in an application for patent as the inventor, or through an error an inventor is not named in an application and such error arose without any deceptive intention on his part, the Commissioner may permit the application to be amended accordingly, under such terms as he prescribes. (Amended August 27, 1982, Public Law 97–247, sec. 6(a), 96 Stat. 320.)

§ 117. Death or incapacity of inventor

Legal representatives of deceased inventors and of those under legal incapacity may make application for patent upon compliance with the requirements and on the same terms and conditions applicable to the inventor.

§ 118. Filing by other than inventor

Whenever an inventor refuses to execute an application for patent, or cannot be found or reached after diligent effort, a person to whom the inventor has assigned or agreed in writing to assign the invention or who otherwise shows sufficient proprietary interest in the matter justifying such action, may make application for patent on behalf of and as agent for the inventor on proof of the pertinent facts and a showing that such action is necessary to preserve the rights of the parties or to prevent irreparable damage; and the Commissioner may grant a patent to such inventor upon such notice to him as the Commissioner deems sufficient, and on compliance with such regulations as he prescribes.

§ 119. Benefit of earlier filing date in foreign country; right of priority

An application for patent for an invention filed in this country by any person who has, or whose legal representatives or assigns have, previously regularly filed an application for a patent for the same invention in a foreign country which affords similar privileges in the case of applications filed in the United States or to citizens of the United States, shall have the same effect as the same application would have if filed in this country on the date on which the application for patent for the same invention was first filed in such foreign country, if the application in this country is filed within twelve months from the earliest date on which such foreign application was filed; but no patent shall be granted on any application for patent for an invention which had been patented or described in a printed publication in any country more than one year before the date of the actual filing of the application in this country, or which had been in public use or on sale in this country more than one year prior to such filing.

No application for patent shall be entitled to this right of priority unless a claim therefor and a certified copy of the original foreign application, specification and drawings upon which it is based are filed in the Patent and Trademark Office before the patent is granted, or at such time during the pendency of the application as required by the Commissioner not earlier than six months after the filing of the application in this country. Such certification shall be made by the patent office of the foreign country in which filed and show the date of the application and of the filing of the specification and other papers. The Commissioner may require a translation of the papers filed if not in the English language and such other information as he deems necessary.

In like manner and subject to the same conditions and requirements, the right provided in this section may be based upon a subsequent regularly filed application in the same foreign country instead of the first filed foreign application, provided that any foreign application filed prior to such subsequent application has been withdrawn, abandoned, or otherwise disposed of, without having been laid open to public inspection and without leaving any rights outstanding, and has not served, nor thereafter shall serve, as a basis for claiming a right of priority.

Applications for inventors' certificates filed in a foreign country in which applicants have a right to apply, at their discretion, either for a patent or for an inventor's certificate shall be treated in this country in the same manner and have the same effect for purpose of the right of priority under this section as applications for patents, subject to the same conditions and requirements of this section as apply to applications for patents, provided such applicants are entitled to the benefits of the Stockholm Revision of the Paris Convention at the time of such filing. (Amended October 3, 1961, Public Law 87–333, sec. 1, 75 Stat. 748; July 28, 1972, Public Law 92–358, sec. 1, 86 Stat. 502; and January 2, 1975, Public Law 93–596, sec. 1, 88 Stat. 1949.)

§ 120. Benefit of earlier filing date in the United States

An application for patent for an invention disclosed in the manner provided by the first paragraph of section 112 of this title in an application previously filed in the United States, or as provided by section 363 of this title, by the same inventor shall have the same effect, as to such invention, as though filed on the date of the prior application, if filed before the patenting or abandonment of or termination of proceedings on the first application or on an application similarly entitled to the benefit of the filing date of the first application and if it contains or is amended to contain a specific reference to the earlier filed application. (Amended November 14, 1975, Public Law 94–131, sec. 9, 89 Stat. 691.)

§ 121.   Divisional applications

If two or more independent and distinct inventions are claimed in one application, the Commissioner may require the application to be restricted to one of the inventions. If the other invention is made the subject of a divisional application which complies with the requirements of section 120 of this title it shall be entitled to the benefit of the filing date of the original application. A patent issuing on an application with respect to which a requirement for restriction under this section has been made, or on an application filed as a result of such a requirement, shall not be used as a reference either in the Patent and Trademark Office or in the courts against a divisional application or against the original application or any patent issued on either of them, if the divisional application is filed before the issuance of the patent on the other application. If a divisional application is directed solely to subject matter described and claimed in the original application as filed, the Commissioner may dispense with signing and execution by the inventor. The validity of a patent shall not be questioned for failure of the Commissioner to require the application to be restricted to one invention. (Amended January 2, 1975, Public Law 93–596, sec. 1, 88 Stat. 1949.)

§ 122.   Confidential status of applications

Applications for patents shall be kept in confidence by the Patent and Trademark Office and no information concerning the same given without authority of the applicant or owner unless necessary to carry out the provisions of any Act of Congress or in such special circumstances as may be determined by the Commissioner. (Amended January 2, 1975, Public Law 93–596, sec. 1, 88 Stat. 1949.)

CHAPTER 12—EXAMINATION OF APPLICATION

Sec.
131. Examination of application.
132. Notice of rejection; reexamination.

§ 131.   Examination of application

The Commissioner shall cause an examination to be made of the application and the alleged new invention; and if on such examination it appears that the applicant is entitled to a patent under the law, the Commissioner shall issue a patent therefor.

§ 132.   Notice of rejection; reexamination

Whenever, on examination, any claim for a patent is rejected, or any objection or requirement made, the Commissioner shall notify the applicant thereof, stating the reasons for such rejection, or objection or requirement, together with such information and references as may be useful in judging of the propriety of continuing the prosecution of his application; and if after receiving such notice, the applicant persists in his claim for a patent, with or without amendment, the application shall be reexamined. No amendment shall introduce new matter into the disclosure of the invention.

§ 133.   Time for prosecuting application

Upon failure of the applicant to prosecute the application within six months after any action therein, of which notice has been given or mailed to the applicant, or within such shorter time, not less than thirty days, as fixed by the Commissioner in such action, the application shall be regarded as abandoned by the parties thereto, unless it be shown to the satisfaction of the Commissioner that such delay was unavoidable.

§ 134.   Appeal to the Board of Appeals

An applicant for a patent, any of whose claims has been twice rejected, may appeal from the decision of the primary examiner to the Board of Appeals, having once paid the fee for such appeal.

§ 135.   Interferences

(a) Whenever an application is made for a patent which, in the opinion of the Commissioner, would interfere with any pending application, or with any unexpired patent, he shall give notice thereof to the applicants, or applicant and patentee, as the case may be. The question of priority of invention shall be determined by a board of patent interferences (consisting of three examiners of interferences) whose decision, if adverse to the claim of an applicant, shall constitute the final refusal by the Patent and Trademark Office of the claims involved, and the Commissioner may issue a patent to the applicant who is adjudged the prior inventor. A final judgment adverse to a patentee from which no appeal or other review has been or can be taken or had shall constitute cancellation of the claims involved from the patent, and notice thereof shall be endorsed on copies of the patent thereafter distributed by the Patent and Trademark Office.

(b) A claim which is the same as, or for the same or substantially the same subject matter as, a claim of an issued patent may not be made in any application unless such a claim is made prior to one year from the date on which the patent was granted.

(c) Any agreement or understanding between parties to an interference, including any collateral agreements referred to therein, made in connection with or in contemplation of the termination of the interference, shall be in writing and a true copy thereof filed in the Patent and Trademark Office before the termination of the interference as between the said parties to the agreement or understanding. If any party filing the same so requests, the copy shall be kept separate from the file of the interference, and made available only to Government agencies on written request, or to any person on a showing of good cause. Failure to file the copy of such agreement or understanding shall render permanently unenforceable such agreement or understanding and any patent of such parties involved in the interference or any patent subsequently issued on any application of such parties so involved. The Commissioner may, however, on a showing of good cause for failure to file within the time prescribed, permit the filing of the agreement or understanding during the six-month period subsequent to the termination of the interference as between the parties to the agreement or understanding.

The Commissioner shall give notice to the parties or their attorneys of record, a reasonable time prior to said termination, of the filing requirement of this section. If the Commissioner gives such notice at a later time, irrespective of the right to file such agreement or understanding within the six-month period on a showing of good cause, the parties may file such agreement or understanding within sixty days of the receipt of such notice.

Any discretionary action of the Commissioner under this subsection shall be reviewable under section 10 of the Administrative Procedure Act. (Amended October 15, 1962, Public Law 87–831, 76 Stat. 958; and January 2, 1975, Public Law 93–596, sec. 1, 88 Stat. 1949.)

Chapter 13—Review of Patent and Trademark Office Decision

§ 141. Appeal to Court of Appeals for the Federal Circuit

An applicant dissatisfied with the decision of the Board of Appeals may appeal to the United States Court of Appeals for the Federal Circuit, thereby waiving his right to proceed under section 145 of this title. A party to an interference dissatisfied with the decision of the board of patent interferences on the question of priority may appeal to the United States Court of Appeals for the Federal Circuit, but such appeal shall be dismissed if any adverse party to such interference, within twenty days after the appellant has filed notice of appeal according to section 142 of this title, files notice with the Commissioner that he elects to have all further proceedings conducted as provided in section 146 of this title. Thereupon the appellant shall have thirty days thereafter within which to file a civil action under section 146, in default of which the decision appealed from shall govern the further proceedings in the case.

§ 142. Notice of appeal

When an appeal is taken to the United States Court of Appeals for the Federal Circuit, the appellant shall give notice thereof to the Commissioner, and shall file in the Patent and Trademark Office his reasons of appeal, specifically set forth in writing, within such time after the date of the decision appealed from, not less than sixty days, as the Commissioner appoints.

§ 143. Proceedings on appeal

The United States Court of Appeals for the Federal Circuit shall, before hearing such appeal, give notice of the time and place of the hearing to the Commissioner and the parties thereto. The Commissioner shall transmit to the court certified copies of all the necessary original papers and evidence in the case specified by the appellant and any additional papers and evidence specified by the appellee and in an ex parte case the Commissioner shall furnish the court with the grounds of the decision of the Patent and Trademark Office, in writing, touching all the points involved by the reasons of appeal.

§ 144. Decision on appeal

The United States Court of Appeals for the Federal Circuit on petition, shall hear and determine such appeal on the evidence produced before the Patent and Trademark Office, and the decision shall be confined to the points set forth in the reasons of appeal. Upon its determination the court shall return to the Commissioner a certificate of its proceedings and decision, which shall be entered of record in the Patent and Trademark Office and govern the further proceedings in the case.

§ 145. Civil action to obtain patent

An applicant dissatisfied with the decision of the Board of Appeals may unless appeal has been taken to the United States Court of Appeals for the Federal Circuit, have remedy by civil action against the Commissioner in the United States District Court for the District of Columbia if commenced within such time after such decision, not less than sixty days, as the Commissioner appoints. The court may adjudge that such applicant is entitled to receive a patent for his invention, as specified in any of his claims involved in the decision of the Board of Appeals, as the facts in the case may appear and such adjudication shall authorize the Commissioner to issue such patent on compliance with the requirements of law. All the expenses of the proceedings shall be paid by the applicant.

§ 146. Civil action in case of interference

Any party to an interference dissatisfied with the decision of the board of patent interferences on the question of priority, may have remedy by civil action, if commenced within such time after such decision, not less than sixty days, as the Commissioner appoints or as provided in section 141 of this title, unless he has appealed to the United States Court of Appeals for the Federal Circuit, and such appeal is pending or has been decided. In such suits the record in the Patent and Trademark Office shall be admitted on motion of either party upon the terms and conditions as to costs, expenses, and the further cross-examination of the witnesses as the court imposes, without prejudice to the right of the parties to take further testimony. The testimony and exhibits of the record in the Patent and Trademark Office when admitted shall have the same effect as if originally taken and produced in the suit.

Such suit may be instituted against the party in interest as shown by the records of the Patent and Trademark Office at the time of the decision complained of, but any party in interest may become a party to the action. If there be adverse parties residing in a plurality of districts not embraced within the same state, or an adverse party residing in a foreign country, the United States District Court for the District of Columbia shall have jurisdiction and may issue summons against the adverse parties directed to the marshal of any district in which any adverse party resides. Summons against adverse parties residing in foreign countries may be served by publication or otherwise as the court directs. The Commissioner shall not be a necessary party but he shall be notified of the filing of the suit by the clerk of the court in which it is filed and shall have the right to intervene. Judgment of the court in favor of the right of an applicant to a patent shall authorize the Commissioner to issue such patent on the filing in the Patent and Trademark Office of a certified copy of the judgment and on compliance with the requirements of law. (Amended January 2, 1975, Public Law 93–596, sec. 1, 88 Stat. 1949; April 2, 1982, Public Law 97–164, sec. 163, 96 Stat. 49.)

CHAPTER 14—ISSUE OF PATENT

Sec.
151. Issue of patent.
152. Issue of patent to assignee.
153. How issued.
154. Contents and term of patent.
155. Patent term extension.
155A. Patent term restoration.

§ 151. Issue of patent

If it appears that applicant is entitled to a patent under the law, a written notice of allowance of the application shall be given or mailed to the applicant. The notice shall specify a sum, constituting the issue fee or a portion thereof, which shall be paid within three months thereafter.

Upon payment of this sum the patent shall issue, but if payment is not timely made, the application shall be regarded as abandoned.

Any remaining balance of the issue fee shall be paid within three months from the sending of a notice thereof and, if not paid, the patent shall lapse at the termination of this three-month period. In calculating the amount of a remaining balance, charges for a page or less may be disregarded.

If any payment required by this section is not timely made, but is submitted with the fee for delayed payment and the delay in payment is shown to have been unavoidable, it may be accepted by the Commissioner as though no abandonment or lapse had ever occurred. (Amended July 24, 1965, Public Law 89–83, secs. 4 and 6, 79 Stat. 260; and January 2, 1975, Public Law 93–601, sec. 3, 88 Stat. 1956.)

§ 152. Issue of patent to assignee

Patents may be granted to the assignee of the inventor of record in the Patent and Trademark Office, upon the application made and the specification sworn to by the inventor, except as otherwise provided in this title. (Amended January 2, 1975, Public Law 93–596, sec. 1, 88 Stat. 1949.)

§ 153. How issued

Patents shall be issued in the name of the United States of America, under the seal of the Patent and Trademark Office, and shall be signed by the Commissioner or have his signature placed thereon and attested by an officer of the Patent and Trademark Office designated by the Commissioner, and shall be recorded in the Patent and Trademark Office. (Amended January 2, 1975, Public Law 93–596, sec. 1, 88 Stat. 1949.)

§ 154. Contents and term of patent

Every patent shall contain a short title of the invention and a grant to the patentee, his heirs or assigns, for the term of seventeen years, subject to the payment of issue fees as provided for in this title, of the right to exclude others from making, using, or selling the invention throughout the United States, referring to the specification for the particulars thereof. A copy of the specification and drawings shall be annexed to the patent and be a part thereof. (Amended July 24, 1965, Public Law 89–83, sec. 5, 79 Stat. 261; December 12, 1980, Public Law 96–517, sec. 4, 94 Stat. 3018.)

APPENDIX C:

Schedule of Patent Office Fees

FEES AND PAYMENT

Following is a list of patent related fees and charges which are payable to the Patent and Trademark Office: *Fees are subject to change in October 1988.*

1. Filing fees.

Basic fee for filing each application of an original patent, except design or plant cases:

By a small entity $170.00

By other than a small entity $340.00

In addition to the basic filing fee in an original application, for filing or later presentation of each independent claim in excess of 3:

By a small entity .... $ 17.00
By other than a small entity .... $ 34.00

In addition to the basic fee in an original application, for filing or later presentation of each claim (whether independent or dependent) in excess of 20:

By a small entity .... $ 6.00
By other than a small entity .... $ 12.00

In addition to the basic fee in an original application, if the application contains, or is amended to contain, a multiple dependent claim(s), per application:

By a small entity .... $ 55.00

By other than a small entity $110.00

Surcharge for filing the basic filing fee or oath or declaration on a date later than the filing date of the application:

By a small entity .... $ 55.00

By other than a small entity $110.00

For filing each design application:

By a small entity .... $ 70.00

By other than a small entity $140.00

Basic fee for filing each plant application:

By a small entity $110.00

By other than a small entity $220.00

Basic fee for filing each reissue application:

By a small entity $170.00

By other than a small entity $340.00

In addition to the basic filing fee in a reissue application, for filing or later presentation of each claim which is in excess of the number of independent claims in the original patent:

By a small entity .... $ 17.00

By other than a small entity .... $ 34.00

In addition to the basic filing fee in a reissue application, for filing or later presentation of each claim (whether independent or dependent) in excess of 20 and also in excess of the number of claims in the original patent:

By a small entity .... $ 6.00

By other than a small entity .... $ 12.00

2. Patent issue fees.

Issue fee for issuing each original or reissue patent, except a design or plant patent:

By a small entity $280.00

By other than a small entity $560.00

Issue fee for issuing a design patent:

By a small entity .... $ 100.00

By other than a small entity $200.00

Issue fee for issuing a plant patent:

By a small entity $140.00

By other than a small entity $280.00

3. Post-issuance fees.

For providing a certificate of correction of applicant's mistake .... $ 29.00

Petition for correction of inventorship in patent $140.00

For filing a request for reexamination $1,770.00

For filing each statutory disclaimer:

By a small entity .... $ 28.00

By other than a small entity .... $ 56.00

For maintaining an original or reissue patent, except a design or plant patent, based on an application filed on or after December 12, 1980 and before August 27, 1982, in force beyond 4 years; the fee is due by three years and six months after the original grant $225.00

For maintaining an original or reissue patent, except a design or plant patent, based on an application filed on or after December 12, 1980 and before August 27, 1982, in force beyond 8 years; the fee is due by seven years and six months after the original grant $445.00

For maintaining an original or reissue patent, except a design or plant patent, based on an application filed on or after December 12, 1980 and before August 27, 1982, in force beyond 12 years; the fee is due by eleven years and six months after the original grant $670.00

For maintaining an original or reissue patent, except a design or plant patent, based on an application filed on or after August 27, 1982 in force beyond 4 years; the fee is due by three years and six months after the original grant:

By a small entity $225.00
By other than a small entity $450.00

For maintaining an original or reissue patent, except a design or plant patent, based on an application filed on or after August 27, 1982, in force beyond 8 years; the fee is due by seven years and six months after the original grant:

By a small entity $445.00
By other than a small entity $890.00

For maintaining an original or reissue patent, except a design or plant patent, based on an application filed on or after August 27, 1982, in force beyond 12 years; the fee is due by eleven years and six months after the original grant:

By a small entity $670.00
By other than a small entity $1,340.00

Surcharge for paying a maintenance fee during the 6-month grace period following the expiration of three years and six months, seven years and six months, and eleven years and six months after the date of the original grant of a patent based on an application filed on or after December 12, 1980 and before August 27, 1982 _____ $110.00

Surcharge for paying a maintenance fee during the 6-month grace period following the expiration of three years and six months, seven years and six months, and eleven years and six months after the date of the original grant of a patent based on an application filed on or after August 27, 1982

By a small entity (§ 1.9(f)) .... $ 55.00
By other than a small entity $110.00

Surcharge for accepting a maintenance fee after expiration of a patent for non-timely payment of a maintenance fee where the delay in payment is shown to the satisfaction of the Commissioner to have been unavoidable $500.00

4. Patent application processing fees.

Extension fee for response within first month:

By a small entity .... $ 26.00
By other than a small entity .... $ 56.00

Extension fee for response within second month:

By a small entity .... $ 85.00
By other than a small entity $170.00

Extension fee for response within third month:

By a small entity $195.00
By other than a small entity $390.00

Extension fee for response within fourth month:

By a small entity $305.00
By other than a small entity $610.00

For filing a notice of appeal from the examiner to the Board of Patent Appeals and Interferences:

By a small entity .... $ 65.00
By other than a small entity $130.00

In addition to the fee for filing a notice of appeal, for filing a brief in support of the appeal:

By a small entity .... $ 65.00
By other than a small entity $130.00

For filing a request for an oral hearing before the Board of Patent Appeals and Interferences:

By a small entity .... $ 55.00
By other than a small entity $110.00

For filing a petition to the Commissioner under a section of this part listed below which refers to 37 CFR 1.17 (h).. $140.00

—for filing by other than all the inventors or a person not the inventor.
—for correction inventorship.
—for decision on questions not specifically provide for.

—to suspend the rules.

—for review of refusal to publish a statutory invention registration.

—for review of decision refusing to accept and record payment of a maintenance fee filed prior to expiration of patent.

—for reconsideration of decision on petition refusing to accept delayed payment of maintenance fee in expired patent.

—for petition in an interference.

—for request for reconsideration of a decision on peition in an interference.

—for late filing of interference settlement agreement.

—for expedited handling of foreign filing license.

—for changing the scope of a license.

—for retroactive license

For filing a petition to the Commissioner under a section of this part listed below which refers to 37 CFR 1.17 (i) ....$ 72.00

—for access to an assignment record
—for access to an application
—for entry of late priority papers
—to make application special
—to suspend action in application
—for divisional reissues to issue separately
—for access to interference settlement agreement
—for amendment after payment of issue fee
—to withdraw an application from issue
—to defer issuance of a patent
—for patent to issue to assignee, assignment recorded late

For filing a petition to institute a public use proceeding $860.00

For processing an application filed with a specification in a non-English language ....$ 26.00

For filing a petition (1) for the revival of an unavoidably abandoned application under 35 U.S.C. 133, or 371, or (2) for delayed payment of the issue fee under 35 U.S.C. 151:

By a small entity ....$ 28.00
By other than a small entity ....$ 56.00

For filing a petition (1) for revival of an unintentionally abandoned application or (2) for the unintentionally delayed payment of the fee for issuing a patent:

By a small entity $280.00
By other than a small entity $560.00

5. International application filing and processing fees.

The following fees and charges are established by the Patent and Trademark Office under the authority of 35 U.S.C. 376:

A transmittal fee..$170.00

A search fee where:

No corresponding prior United States national application with fee has been filed $420.00

Corresponding prior United States national application with fee has been filed $250.00

Search fee with the European Patent Office as the International Searching Authority $930.00*

A supplemental search fee when required by the United States Patent and Trademark Office..$140.00

(Any supplemental search fee required by the European Patent Office must be paid directly to that Office.).

The national fee, that is, the amount set forth as the filing fee under filing fees above credited if requested at time of filing by an amount of $170.00 where an international search fee has been paid on the corresponding international application to the United States Patent and Trademark Office as an International Searching Authority.

Surcharge for filing the patents fee or oath or declaration later than 20 months from the priority date:

By a small entity .... $ 55.00

By other than a small entity $110.00

For filing an English translation of an international application later than 20 months after the priority date .... $ 26.00

International Fees Basic Fee (first 30 pages) $375.00*

Basic Supplemental Fee (for each sheet over 30) .... $ 7.00*

Designation Fee for the first 10 (per national or regional offices) .... $ 90.00*

Designation Fee for the 11th and subsequent designations .... No charge

Note: PCT fees controlled by WIPO & EPO will fluctuate with exchange rates.

6. Service fees

The Patent and Trademark Office will provide the following services upon payment of the fees indicated:

Printed copy of a patent, including a design patent, statutory invention registration, or defensive publication document, except color plant patent or color statutory invention registration .... $ 1.50

Printed copy of a plant patent or statutory invention registration in color .... $ 6.00

Copy of patent application as filed .... $ 9.00

Copy of patent file wrapper and contents, per 200 pages or a fraction thereof .... $ 75.00

Copy of Office records, except as otherwise provided in this section, per page .... $ .50

Microfiche copy of microfiche, per microfiche .... $ .50

For certifying Office records, per certificate .... $ 3.00

For a search of assignment records, abstract of title and certification, per patent .... $ 12.00

To compare and certify copies made from Patent and Trademark Office records but not prepared by the Patent and Trademark Office, per copy of document .... $ 5.00

Subscription services:

(1) Subscription orders for printed copies of patents as issued, annual service charge for entry of order and ten subclasses .... $ 7.00

(2) For annual subscription to each additional subclass in addition to the ten covered by the fee under (1) above, per subclass .... $ .70

For providing to libraries copies of all patents issued annually, per annum .... $ 50.00

For list of all United States patents and statutory invention registrations in a subclass, per 100 numbers or fraction thereof .... $ 1.00

For list of United States patents and statutory invention registrations in a subclass limited by date or number, per 50 numbers or fraction thereof .... $ 1.00

Disclosure document: For filing a disclosure docuemtn .... $ 6.00

Search of Office records: For searching Patent and Trademark Office records for purposes not otherwise specified, per one-half hour or fraction thereof .... $ 14.00

Recording of documents:

(1) For recording each assignment, agreement or other paper relating to the property in a patent or application .... $ 7.00

(2) Where a document to be recorded under paragraph (1) above refers to more than one patent or application, for each additional patent or application .... $ 2.00

Publication in Official Gazette: For publication in the Official Gazette of a notice of the availability of an application or a patent for licensing or sale, each application or patent .... $ 7.00

For a duplicate or replacement of a permanent Office user pass. (There is no charge for the first permanent user pass) .... $ 5.00

Delivery box: Local delivery box rental, per annum .... $ 43.00

CopiShare card: Cost per copy .... $ .20

International-type search reports: For preparing an international-type search report of an international-type search made at the time of the first action on the merits in a national patent application .... $ 28.00

Microfiche copy of patent file record .... $ 6.00

Uncertified statement as to status of the payment of maintenance fees due on a patent or expiration of a patent... $ 3.00

For processing and retaining any application abandoned pursuant to §1.53(d) unless the required basic filing fee has been paid $100.00

Handling fee for application filed without the specification or drawing required by §1.53(b) and the omission is not corrected within the time period set .... $ 15.00

Handling fee for withdrawal of Statutory Invention Registration............................... $100.00

Uncertified copy of a non-United States patent document, per document .... $ 10.00

Copy of patent assignment record .... $ 1.50

Additional filing receipts

—Duplicate.. $ 14.00
—Corrected due to applicant error .... $ 14.00

7. Miscellaneous fees and charges.

The Patent and Trademark Office has established the following fees for the services indicated:

Deposit accounts:

(1) For establishing or reinstating a deposit account .... $ 8.00
(2) Service charge for each month when the balance at the end of the month is below $1,000..$ 20.00
(3) Service charge for each month when the balance at the end of the month is below $300 for restricted subscription deposit accounts used exclusively for subscription order of patent copies as issued ....$ 20.00

Registration of attorneys and agents:

(1) For admission to examination for registration to practice, fee payable upon application $250.00
(2) On registration to practice... $ 81.00
(3) For reinstatement to practice .... $ 9.00
(4) For certificate of good standing as an attorney or agent.................................... $ 10.00
    Suitable for framing...$ 88.00
(5) For review of a decision of the Director of Enrollment and Discipline under §10.2(c).$ 92.00
(6) For requesting regrading of an examination under §10.7(c)................................ $ 92.00
For processing each check returned "unpaid" by a bank.. $ 20.00

Note: PCT fees controlled by WIPO & EPO will fluctuate with exchange rates.

The following publications are sold, and the prices for them fixed, by the Superintendent of Documents, Government Printing Office, Washington, D.C. 20402, to whom all communications respecting the same should be addressed:

Official Gazette of the United States Patent Office:
Annual subscription, domestic (non-priority) $270.00
Annual subscription, (priority) $375.00
Annual subscription, foreign international Postal Zone... 337.50
Single numbers 13.00
Annual Index Relating to Patents, price varies
Manual of Classifications of Patents 77.00
    Foreign... 89.25
Attorneys and Agents Registered to Practice Before the U.S. Patent Office...................... 17.00
37.Code of Federal Regulations 8.00
Manual of Patent Examining Procedure 70.00
    Foreign... 87.50

THE ABOVE PRICES ARE SUBJECT TO CHANGE WITHOUT NOTICE.

All payment of money required for Patent and Trademark Office fees should be made in United States specie, Treasury notes, national bank notes, post office money orders or postal notes payable to the Commissioner of Patents and Trademarks, or by certified checks. If sent in any other form, the Office may delay or cancel the credit until collection is made. Postage stamps are not acceptable. Money orders and checks must be made payable to the Commissioner of Patents and Trademarks. Remittances from foreign countries must be payable and immediately negotiable in the United States for the full amount of the fee required. Money paid by actual mistake or in excess, such as a payment not required by law, will be refunded, but a mere change of purpose after the payment of money, as when a party desires to withdraw his application for a patent or to withdraw an appeal, will not entitle a party to demand such a return. Amounts of $1.00 or less will not be returned unless specifically demanded, within a reasonable time.

APPENDIX D:

Recommended Reading

U.S. GOVERNMENT PUBLICATIONS

Manual of Patent Office Examining Procedure (M.P.E.P.)
This is the best reference for detailed information regarding the patent application process. It is the standard reference manual used by Patent Office staff members. It can be purchased from the Government Printing Office in Washington, D.C. Contact your local federal book store for ordering information.

Patent Laws
This is a compilation of acts of Congress pertaining to patents. It is carried in most federal book stores.

Code of Federal Regulations, Title 37
This is the book of Patent Office "rules." The rules are official regulations promulgated by the Department of Commerce. 1 They have the force of law. This is also available at local federal book stores.

TEXT BOOKS

Kayton, *Patent Prosecution and Practice* (Patent Resources Institute, Inc., Washington. D.C.)
This is an excellent, practical, multi-volume guide to patent application writing. It is available at major law libraries.

Landis, *Mechanics of Patent Claim Drafting* (Practicing Law Institute, New York City)
This book provides helpful hints concerning claim writing.

Index